Breeding Contempt

Breeding Contempt

The History of Coerced Sterilization in the United States

Mark A. Largent

RUTGERS UNIVERSITY PRESS
NEW BRUNSWICK, NEW JERSEY, AND LONDON

Library of Congress Cataloging-in-Publication Data

Largent, Mark A.

Breeding contempt : the history of coerced sterilization in the United States / Mark A. Largent.

p. cm.

Includes bibliographical references and index.

ISBN 978–0–8135–4182–2 (hardcover : alk. paper)—ISBN 978–0–8135–4183–9 (pbk. : alk. paper)

1. Involuntary sterilization—United States—History.
2. Eugenics—United States—History. I. Title.

HV4989.L275 2007

363.9'7—dc22

2007000029

A British Cataloging-in-Publication record for this book is available from the British Library.

Visit our Web site: http://rutgerspress.rutgers.edu

Manufactured in the United States of America

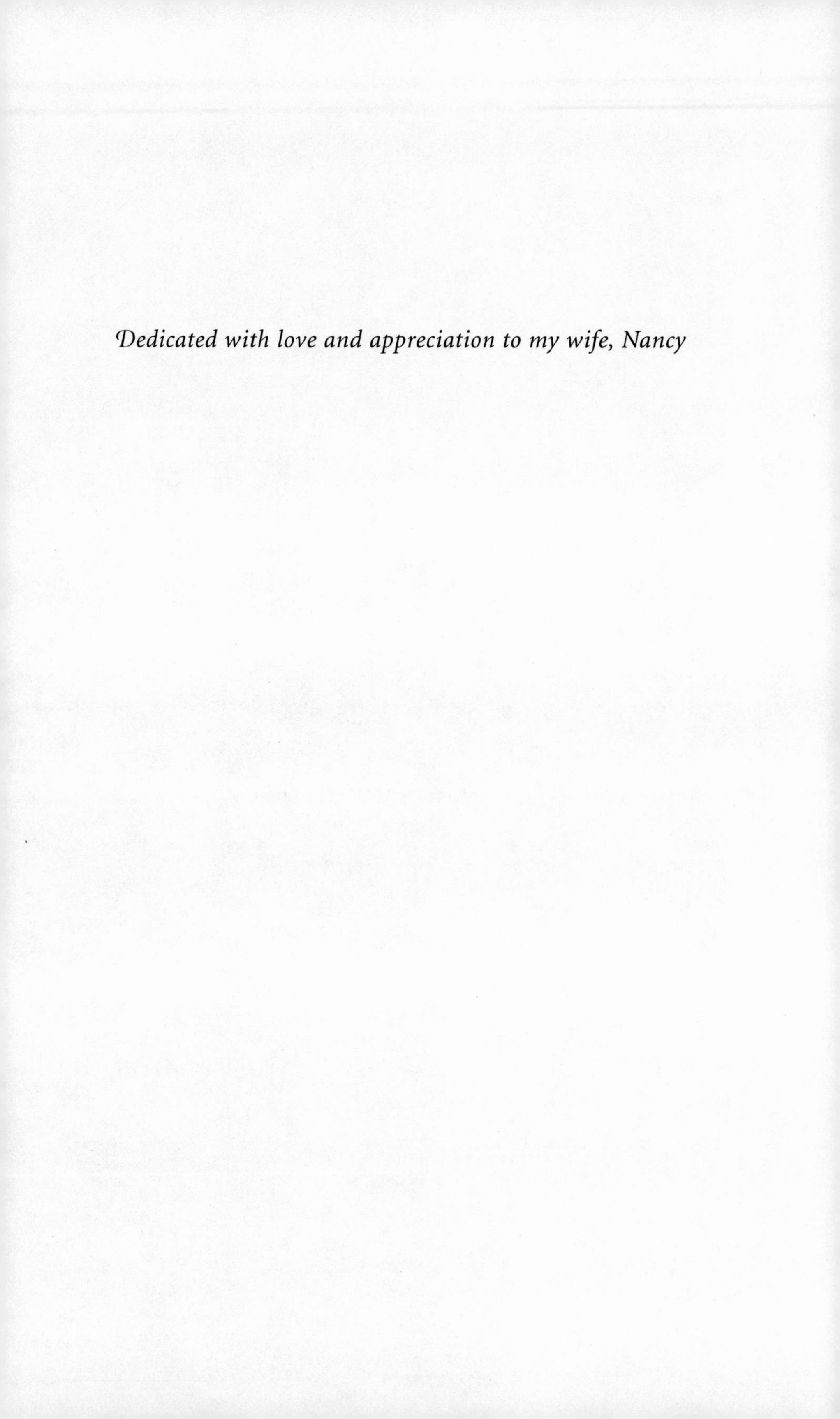

Dedicated with love and appreciation to my wife, Nancy

Contents

Acknowledgments

I owe a tremendous debt of thanks to the many people who have helped me in my career and in writing this book. My parents, Frank and Betty Largent, encouraged me to pursue my interests wherever they might lead me. My advisers, Andrew Conteh at Moorhead State University, Constance Hilliard at the University of North Texas, and especially Sally Gregory Kohlstedt at the University of Minnesota, all took a personal and sincere interest in my success. I would not have completed any of my degrees without their attention and concern. Paul Farber, Mott Greene, and Keith Benson have been my mentors since I left graduate school. Together they helped me think about how my work fit into other scholarship and showed me how to make the transition from student to professor.

Throughout my education and early career I have had the great fortune of becoming friends with a number of amazing scholars. My colleagues at the University of Minnesota's Program in the History of Science; Oregon State University's History Department; the University of Puget Sound's History Department and its Science, Technology and Society Program; and most recently at James Madison College at Michigan State University encouraged my work on this subject. Among those who I want to single out for special thanks is John P. Jackson, who was the first to read an entire draft of the book and whose many comments and additions significantly improved it. John is the smartest person I know, a brilliant scholar and a good friend. Jeff Matthews and Chris Young have been longtime friends, and they both helped me complete this project. Chris listened to long, convoluted explanations of what I thought I was trying to say, and he provided me a title that helped clarify it all. Jeff took the time to read and comment on the manuscript on short notice, and his suggestions substantially improved it. Other friends and colleagues who deserve special thanks include Erik Conway, Steve Fifield, Eric Berg, Kai-Henrik Barth, Ross Emmett, David Sepkoski, Jennifer Gunn, Michael Reidy, Matt Zierler, Dan Kramer, Greg Elliott, Julia Burba, Kristin Johnson, Rich Bellon, Nick Kontogeorgopoulos, Chris Foley, and Doug Sackman. I also want to thank my students, many of whom I count among my best critics and collaborators. In particular, Christine Manganaro, Frazier Benya, and Jasmine Garamella, all of whom were my students at the University of Puget Sound, were particularly helpful as I gathered material for this book. At Michigan State University, two professorial assistants, Jessica Ports and Brittany Foley, tracked down and copied most of the source material for this book, and Jessica suffered through a final proofread of the entire manuscript. Other students, like Emily Carlson, helped by babysitting so that I could finish the manuscript. I also owe a debt of gratitude to the American Philosophical Society, especially Rob Cox, and to all the historians and archivists there who have created a tremendous resource for those interested in the history of the American eugenics movement.

My work seeks to build on what dozens of other excellent scholars have already contributed to our understanding of the history of the American eugenics movement and of coerced sterilization in the United States. There are several whose work has been particularly influential on my scholarship, including Paul Lombardo, Diane Paul, and Gar Allen, and they deserve special recognition. I would also like to thank Leila Zenderland, who encouraged this book throughout its development. Her suggestions provided valuable direction at the project's inception and at its conclusion. Any shortcomings in the book are, of course, my responsibility.

Finally, and most significantly, I want to thank my wife, Nancy, to whom I dedicate this book. Nancy and I met shortly before I started graduate school, and she has been a constant source of inspiration and encouragement throughout the fifteen years we have been together. She supported me through my education, she has again and again listened to me ramble on about long-dead biologists, and she has pushed me to think carefully about my research and the contributions my work can make to the world. She has also given me the greatest gift of all, our daughter, Annabelle. Thank you, Nancy.

Breeding Contempt

Introduction

In the Name of Progress

American physicians coercively sterilized tens of thousands of their patients over the last 150 years. Their efforts began around 1850, and by the 1890s the movement had grown into a full-blown crusade to sterilize or asexualize people who doctors believed would produce undesirable children. Even though they exerted significant influence on American culture, physicians alone could not garner the public support and ultimately the legislation necessary to allow them to coercively sterilize the unfit. Shortly after the turn of the century, several other groups of professionals joined them, including biologists, social scientists, and lawyers. Within four decades, two-thirds of the states had enacted laws that required the sterilization of various criminals, mental health patients, epileptics, and syphilitics. By the early 1960s, more than 63,000 Americans were coercively sterilized under the authority of these laws.

What is known about the practice of compulsory sterilization in the United States has generally been examined as part of the broader story of the American eugenics movement, which has received considerable attention by historians, cultural study scholars, journalists, and, on occasion, social and natural scientists.[1] Most closely associated with the Nazis and World War II atrocities, eugenics is sometimes described as a government-orchestrated breeding program, other times as a pseudoscience, and often as the first step down a slippery slope that inevitably leads to genocide. By the end of the twentieth century, the word *eugenics* had become a slur, something to be avoided at all costs. Occasionally, though, we still see attempts to resurrect the eugenics movement, such as Richard Lynn's 2001 *Eugenics: A Reassessment*, Nicholas Agar's 2004 *Liberal Eugenics: In Defense of Human Enhancement*, or John Glad's 2006 *Future Human Evolution: Eugenics in the Twenty-first Century*, but these books represent the extremes in a conversation that typically depicts eugenics as deeply problematic.[2]

Histories of coerced sterilization in the United States emerged in the 1960s, and they placed responsibility for the movement on a few select men who had participated in the American eugenics movement. Charles Benedict Davenport and his employee at the Eugenics Record Office (ERO), Harry Hamilton Laughlin, are universally identified as the leaders of the American eugenics movement and are often held personally responsible for the sterilization movement. Davenport helped organize a broad, loose collection of sterilization advocates, and his work provided them a scientific legitimacy. His claims about the biological basis of American degeneracy were supported by the research of other American biologists; in turn, Davenport provided these biologists with research facilities and funding. Davenport's authority among his colleagues was based on his ability to raise large sums of money from philanthropists and to organize scientists, who were notoriously inept at organizing themselves. His public authority was founded on his notable scientific credentials and his willingness to offer stark and often startling conclusions based on limited scientific research. His claims simultaneously motivated the public's concern about American citizens' genetic failings and encouraged even more contributions for basic scientific research that patrons hoped would uncover problems and ultimately produce solutions for them. Davenport's protégé, Harry Laughlin, was the chief proponent of compulsory sterilization laws in the United States, and he wrote books and articles on the subject and testified before legislators and judges. The two men provided scientific authority to support the prejudices that had motivated earlier advocates of the coerced sterilization of state wards.

Assigning responsibility for something as vast and influential as the American eugenics movement to one man or even a small group of men is a vestige of what historians ruefully call "great man history." This approach chronicles the wondrous achievements of the select few who steered our ships of state, led our armies, and shaped our cultures. By the latter half of the twentieth century, such histories came to be seen as passé, even dangerous according to some, because they inappropriately and heroically credited all of society's achievements to the work of a few significant figures. Such narrow approaches to the past have since been replaced with social histories, group biographies, and narratives that emphasize cultural trends or the impact of particular events in shaping history. The history of eugenics and the troubled history of coerced sterilization in the United States are just now beginning to move away from "great man history" to describe the roles that many scientists, physicians, and other social authorities played in advancing it. Support for coerced sterilization was widespread, and actually still remains well supported in many corners of American society. The sterilization movement was not isolated to a few places, it was not an aberration, and it did not disappear after World War II. In short, Davenport and Laughlin did not do it alone. In fact, it is difficult to find many early-twentieth-century American biologists who were not advocates of eugenics in some form or another.

Even extending responsibility beyond Davenport and Laughlin to include most American biologists still does not adequately identify all those responsible for the tens of thousands of coerced sterilizations. American biologists were merely one segment of a movement that included thousands of scientists, politicians, social activists, philanthropists, educators, and assorted do-gooders, racists, and utopians. Moreover, many biologists who supported eugenics actually opposed compulsory sterilization, especially when it was performed to allow patients and inmates to reintegrate into society. The public claims made by Davenport and his colleagues most certainly helped motivate the wave of state legislation from 1907 to 1937 that empowered American physicians to sterilize patients with or without their permission. But eugenicists were as much influenced by the cultures in which they operated as they were capable of influencing it. To understand the history of coerced sterilization in the United States, we need to understand the relationships between the leaders of the movement and their cultural, intellectual, and professional contexts. The influences that brought about compulsory sterilization laws existed separately from the individuals who advocated them, and they continued long after the passing of Davenport and his generation of biologists.

One need not be an apologist to admit that the majority of sterilization advocates were well-meaning professionals who saw in the operation viable solutions to complex social problems and devastating physical ailments. It is easy to be presentist about eugenics and to dismiss it as the product of quacks and racists. However, in an age before antibiotics and insulin, before Western medicine's therapeutic capacities began to emerge, preventing the birth of individuals encumbered with genetic ailments seemed the only humane and prudent cure to many of the professionals described in this book. Davenport and his colleagues operated at the time when medical doctors could do little more than comfort patients, amputate infected limbs, and augment the body's natural healing capabilities. Recognizing that a number of ailments, such as hemophilia, color blindness, and Huntington's chorea, were passed genetically from one generation to another, sterilization proponents seized on the opportunity to prevent a great deal of suffering, expense, and social upheaval, while at the same time furthering their professional status. But most advocates of compulsory sterilization did not stop with genetic diseases; they also imagined that certain behaviors, such as what they considered sexual perversions or criminality, were likewise passed from parent to child either through genetic or cultural inheritance. Laughlin, for example, believed that by sterilizing "socially inadequate citizens," authorities could solve any number of complicated social problems.[3]

This is a story about how good intentions and professional authority can produce horrible results. From the point of view of the twenty-first century, it is a story that shows how American professionals dehumanized large groups of people—be they gay, minority, physically or mentally ill, poor, criminalistic, sexually deviant, or mentally challenged—and treated them in incredibly brutal ways. It is a story about the hubris of professionals both yesterday and today. It is, quite honestly, an attempt to spread the responsibility for the sterilization of tens of thousands of mental health patients, welfare recipients, and prisoners off the shoulders of a few select people. This book does not apologize for the eugenicists, nor does it advocate a renewal of the eugenics movement. The purpose of this story is to demonstrate the power of professionals in American culture and to lay bare our ongoing compulsion to locate the source of complex problems within certain citizens' testicles and ovaries.

A Word about Terminology

The terminology that medical providers, politicians, biologists, and social scientists used in describing the mentally deficient, mentally ill, felons, homosexuals, and the poor is distasteful and stigmatizing. Yet terms such as *sexually and morally perverted*, *degenerate*, *feebleminded*, *retarded*, *inferior*, and *unfit* are employed in this book not for shock value or because they are accurate and appropriate descriptions. Rather, the use of archaic and often disturbing language is a recognition that such words signified specific assumptions. Sanitizing their language by translating offensive terms into contemporary labels would do much more than merely lose the flavor of their discourse, it would camouflage some of the assumptions under which professionals labored. In addition, much of our currently accepted language includes euphemisms that gloss over unpleasant realities or particular difficulties in some citizens' lives. In these cases, the earlier language offers more candid descriptions of their challenges. Sometimes, though, earlier vocabulary, such as "the crime against nature," similarly obscures their original meanings. In these cases, I will explain as best I can precisely what users of the language meant in their adoption of particular terms.[4]

We have not overcome the legacy to which earlier, harsher words alluded. Cleansing our language does not necessarily purge underlying assumptions, but it certainly does obscure them. The power of direct language, devoid of the jargon on which professionals rely, was most powerfully demonstrated to me in an encounter with a gay rights activist early in my research on the subject of coerced sterilization. After I learned about his interest in Oregon's compulsory sterilization laws, we met at a coffee shop in downtown Portland. A well-groomed, fit-looking man in his early fifties and dressed in an expensive suit, he showed me a stack of photocopies of records relating to sterilized prisoners. In the middle of a busy coffee shop, he paged through them and described in graphic language the crimes that the men had been convicted of committing. The locker-room language he employed in discussing activities that are more politely and euphemistically described as "sodomy" or "the crime against nature" stood in stark contrast to his professional appearance. It left me confused about his intentions and embarrassed by the fact that his booming voice was easily overheard by everyone in the coffee shop. Later, I asked a scholar who works on the history of sexuality about the incident. He explained that some activists eschew contemporary euphemisms because they obscure and thus demonize the activities the words weakly describe. Instead, many activists employ graphic descriptions of sexual acts that leave no room for prudish denigration of the sex lives of consenting adults. To some degree, I have adopted this approach and use terms throughout the book that reflect the assumptions and words used by historical actors to portray as vividly as possible their worldviews.

In general, I use the term *sexual surgeries* to refer to the procedures that purposefully altered patients' and prisoners' sexual abilities, urges, or reproductive potentials. The operations aimed at fundamentally restricting patients' abilities to procreate and in some cases even to participate in sexual activities. Whereas most other surgeries, elective or otherwise, are intended to augment a person's physical abilities or preserve their life, the sexual surgeries on which I focus here generally reduce patients' abilities to engage in sexual activities. In the case of truly voluntary sexual surgeries, the limitations that the operations create are desired by the patients and enhance their qualities of life. However, with coerced sterilizations, sexual surgeries eliminated patients' abilities to procreate and in some cases engage in sexual intercourse, often to the great regret of the targeted populations.

From the mid-nineteenth century through the late twentieth century, medical authorities used a number of methods to asexualize or sterilize patients and a large number of terms to describe the procedures. There were two different classes of operations: the complete removal of sex organs and the more limited surgical procedures that rendered some or all of an individual's sex organs inoperative. When used by medical doctors, the term *castration* referred generally to removal of the testes from a man or the ovaries from a women; in legislative and popular discourse, it was almost always called emasculation or asexualization. In the case of men, all of these terms referred to the removal of the testes or amputation of the testes and scrotum. Among professionals, the amputation of the testes and scrotum was formally described as an orchidectomy or a testiectomy, dramatic surgeries that were used from the late nineteenth century through the 1930s to treat, punish, or control hundreds of rapists, child molesters, and men who engaged in activities associated with homosexuality. There was some discussion at the turn of the century of an operation called a spermectomy, in which the cordlike structure that contained the vas deferens and its accompanying arteries, veins, nerves, and lymphatic vessels was severed. Both the spermectomy and the orchidectomy would eventually eliminate the ability of a man to have an erection or to ejaculate. This surgery was seldom performed in the context of eugenic sterilization, but it was advocated throughout the 1890s by those who feared that merely sterilizing defectives, especially rapists, would encourage increased deviant behavior. The most common sexual surgery in men was the vasectomy, which was first used near the end of the nineteenth century and popularized as a method to limit fertility without hindering the patient's ability to engage in sexual activity. Quite different from castration, vasectomies were sometimes advertised as capable of increasing men's vigor, as was the case with the work done by Eugen Steinach in the early twentieth century.[5] In the 1980s, a new form of castration, chemical castration, emerged and became popular as both a punishment and a treatment for men convicted of sexual offenses. In women, there were also two categories of sexual surgeries, a dramatic intervention that removed portions of the patient's reproductive organs and a more limited operation that destroyed her power to procreate. Advocates of compulsory sterilization who called for the asexualization or castration of women generally wanted the woman to undergo a hysterectomy, which consisted of the removal of the uterus, or an ovarioectomy, removal of the ovaries, sometimes called an oöphorectomy. Another less dramatic surgery involved severing or removing the fallopian tubes, which was called a salpingectomy, tubal ligation, fallectomy, tuberotomy, or tuberectomy. Similar to men, the more dramatic surgeries in women resulted in hormonal changes, while simple sterilization operations merely destroyed the patient's ability to have children.

I use the term *coerced sterilization* in reference to the general movement among American professionals to use sexual surgeries to solve complex and vexing social, economic, and moral dilemmas. Notions of what was voluntary or involuntary changed considerably throughout the twentieth century, especially with the emergence of the Nuremberg Code in 1947 and the evolution of informed consent as a vital concept in American medical care. Compulsory sterilization laws—the state-by-state legal framework that brought about the coerced sterilization of prisoners, mental health patients, and other wards of the state—first began to be debated in the mid-nineteenth century. Coercion, on the other hand, could take many forms—legal, economic, or social—and in some cases patients were coerced to "volunteer" for the surgeries.[6] In discussing laws that mandated the use of sterilization or asexualization by prison officials, asylum managers, physicians, and mental health care providers, I use the term *compulsory sterilization* for two reasons. First, it is the term generally used by twentieth-century advocates of legislation who employed sterilization as a tool of social and biological improvement. Second, it calls attention to the fact that sterilization laws allowed the state to use the full range of police powers in compelling individuals to relinquish their reproductive abilities.

There were three fundamentally different eras in which American physicians employed sexual surgeries, either asexualization or simple sterilization. In the first, which began in the late nineteenth century and lasted until the turn of the twentieth, the prevention of crime was the key motivating factor, and medical doctors were by far the most aggressive proponents. Castration was the usual remedy, and advocates used punitive and therapeutic justifications when they operated on deviants. Unlike the claims made in previous histories of the American eugenics movement or of coerced sterilization in the United States, these early coerced sterilizations were not done "in the name of eugenics."[7] Instead, physicians offered punitive or therapeutic justifications for the surgeries. Crime was often discussed in relation to a number of sexual perversions that were collectively identified as a crime against nature or sodomy, which included bestiality, anal intercourse, pederasty, and oral sex, and the laws were used to prosecute homosexuals, rapists, and pedophiliacs.

The second era of sexual surgeries began shortly after 1900, when the American eugenics movement emerged and exerted increasing influence on public discussions about the use of sexual surgeries to improve Americans' overall genetic quality.

During this stage, American biologists joined physicians in advocating the use of less intrusive sterilization operations to treat patients they believed were incapable of responsibly using their procreative powers and to eugenically improve the nation's population. Biologists offered the movement scientific authority and the apparent ability to prove physicians' anecdotes about how successive generations of degenerates were in fact caused by a family's tainted heredity. Compulsory sterilization attained its highest level of popularity in the 1930s. In 1937, for example, *Fortune* magazine asked in its annual readers' survey about sterilization: "Some people advocate compulsory sterilization of habitual criminals and mental defectives so that they will not have children to inherit their weaknesses. Would you approve of this?" Sixty-six percent of the respondents believed that mental defectives should be sterilized, and 63 percent favored sterilizing criminals. Less than one in six respondents directly opposed sterilization.[8]

The third and final era in the history of coerced sterilization began shortly after midcentury, when increasing numbers of people demonstrated their resentment to government's intrusion into citizens' reproductive lives and several professions abandoned the movement. In the 1960s and 1970s, as older eugenic rationales came under attack by activists, government coercion was replaced by the potential of new tools—like amniocentesis, genetic screening, and legalized abortion—to empower parents to make informed decisions about their offspring. Shortly thereafter, the punitive justifications present throughout the first era reemerged. In the 1980s, legislatures and courts again began employing sexual surgeries to punish welfare recipients, rapists, and child molesters. As was the case a hundred years earlier, many government officials believed that the source of complex social problems was located within the genitals of certain citizens. Advocates of sexual surgeries in the twenty-first century extend from district attorneys' offices to the highest levels of the federal government; when he was governor of Texas, President George W. Bush publicly supported the castration of a convicted child molester to curb his degenerate impulses.[9]

Coerced Sterilizations in the United States

As a percentage of the overall population, the number of Americans who were coercively sterilized in the twentieth century was small. California was single-handedly responsible for at least a quarter of the nation's coerced sterilizations, yet 1942, the state's most active year, witnessed the sterilization of only 1 in every 5,800 citizens.[10] But viewed in aggregate terms, the number of Americans who lost their reproductive capacities under the coercion of state laws and institutional mandates is stunning: more than 63,000 people. Coercive sterilization laws persisted longer than any epidemic, and they disproportionately affected individuals in a few particular states. California, Oregon, North Dakota, Kansas, Delaware, and Virginia were especially aggressive in implementing their sterilization laws. Some states' doctors, like those in Illinois, Pennsylvania, and Kansas, did not wait for the passage of compulsory sterilization laws. They sterilized patients on their own authority, which confounds both our notions of the importance of compulsory sterilization laws and our ability to accurately tally the total number of coerced sterilizations in the United States.

In the history of coerced sterilization, several locales loom large. Physicians in Texas were the earliest advocates of coerced sterilization. Michigan was the first state to seriously consider legislation that would compel some of its citizens to be sterilized. Medical doctors, lawyers, judges, and scientists in Chicago were also early advocates of sterilization, and between 1910 and 1930 they were among the most ardent proponents of eugenic sterilization. California, by far, sterilized more of its citizens than did any other state. In sum, over two-thirds of American states adopted sterilization laws in the first four decades of the twentieth century. The widely quoted number of 63,000 is probably not even close to the actual number of coerced sterilizations because so many went unreported, occurred in states that had no legal oversight of coerced sterilization, or were wrongly reported as voluntary when in fact the patient or inmate was coerced by prison authorities or health officials.

Contrary to conventional wisdom, the American eugenics movement did not end with the discovery of the Nazis' atrocities during World War II, nor did the practice of coerced sterilization. From the late 1940s through the 1950s, eugenically justified legislation to limit immigration persisted, hospitals and prisons coercively sterilized record numbers of Americans, and educators in high school and college biology courses continued to teach students how "a great reduction in human suffering could be achieved" if it were possible to "decrease the number of afflicted individuals born in each generation."[11] Far from receding, the compulsory sterilizations continued throughout the postwar years, riding a wave of enthusiasm for science and technology and benefiting from the goodwill most Americans felt toward the medical community as a result of wartime advances in research and public health.

Legal scholars and social scientists finally shifted their position on compulsory sterilization in the late 1950s and early 1960s. It took another decade before Davenport's successors, the next generation of American biologists, abandoned their support for eugenics and coerced sterilization. Between 1968 and 1974, American biologists began attacking eugenics as scientifically problematic and socially unacceptable. In the 1970s, the writers of American biology textbooks increasingly linked the American eugenics movement with the atrocities committed by the Nazis during World War II. Encouraged by an array of civil rights movements—including the disability rights, patient rights, prisoner rights, American Indian, and women's movements, as well as demands from African Americans for social and legal reforms—in the 1970s American biologists took a strong position against eugenics and coerced sterilization.

By the early 1980s, the American eugenics movement was officially dead, and the word *eugenics* had itself acquired derogatory connotations. Nowhere was this transformation more obvious than in the definitions for the word *eugenic* in the 1993 edition of the *Oxford English Dictionary* (*OED*). In the examples that it provided from

1916, 1934, and 1957, one of which referenced Davenport, the word was used simply to describe anyone interested in the genetic quality of a given population. But in the 1989 example, the word *eugenic* was clearly a slur. The *OED* quoted an article from *Atlantic* magazine: "We hope that Professor Herrnstein's eugenicist jeremiad did not make readers too anxious about America's future."[12] Americans obviously changed what they meant by the word *eugenic* sometime between 1957 and 1989, but its significance for the use of sexual surgeries in the twenty-first century is still unclear. As long as we continue to imagine that the coerced sterilization of tens of thousands of Americans was the responsibility of only a small number of evildoers, all of whom are now dead, and that the scientific errors that supposedly brought about the compulsory sterilization movement have been remedied, we are at serious risk of repeating grave mistakes. As genetic technologies grow increasingly powerful, prospective parents are ever-enthusiastic about the ability of modern medicine to help them produce healthy children, and political leaders continue to search for easy solutions to complex social and economic problems, the assumptions on which the compulsory sterilization movement was founded will continue to threaten some Americans' basic civil liberties.

Breeding Contempt

The organization of this book follows the history of coerced sterilization in the United States from the middle of the nineteenth century through the turn of the twenty-first. For well over a hundred years, American legal, scientific, and medical professionals advocated the sterilization of certain Americans to punish them, to treat them for both real and perceived ailments, and to prevent them from passing their traits to the next generation. The story is presented chronologically, beginning with a Texas physician's failed campaign to pass a compulsory sterilization law in the 1850s and ending with current debates about the use of chemical castration as a treatment and a punishment for sex criminals and as a way to control the fecundity of welfare recipients.

The first chapter focuses nearly exclusively on American physicians and their advocacy of sterilization to prevent crime and punish criminals. While a great deal of attention has been given to American biologists and their advocacy of eugenics, very little thought has been paid to the dozens of prominent medical doctors, organizations, and journals that advocated coerced sterilization years before any biologist took up the cause. American physicians were, in fact, pioneers in the movement, and because they developed and performed the sterilization operations, they were vital to its practice.

Despite being overstated in the histories of eugenics, biologists did play a critical role in the emergence of compulsory sterilization laws in the twentieth century. Chapter 2 explores their influence and the rewards they reaped for advocating coerced sterilizations and participating in the American eugenics movement. In the 1880s and 1890s, physicians were using anecdotal evidence to support their claim that human deficiencies had a genetic basis. The rediscovery of Mendel's work in 1900 and the creation of the Station for Experimental Evolution, the American Breeders Association, and the ERO provided advocates of compulsory sterilization laws with the scientific justifications needed to claim that social and physical ailments had a hereditary basis.

The book's third chapter analyzes the emergence, influence, and eventual demise of compulsory sterilization laws in the United States. Histories of coerced sterilization have, because of the localized nature of the subject, tended to examine single states or regions, while histories of the American eugenics movement often gloss over local differences and make broad claims about its effects. Chapter 3 summarizes both the legislation in various states that allowed physicians to coercively sterilize state wards and the court challenges to those laws.

The last two chapters explore the gradual emergence of professional resistance to the use of sexual surgeries for eugenic, therapeutic, and punitive purposes. Public health and the oversight of allied health professionals in the United States are both constitutionally allocated to the states, rather than to the federal government. While there is a general American eugenics movement and nationwide professional organizations, coercive sterilization was in practice a decentralized activity. It was controlled by state authorities and, more often than not, by institutional authorities. Throughout the years between 1910 and 1930, opposition to compulsory sterilization laws was likewise decentralized and largely ineffective. This changed in 1927 with the release of Oliver Wendell Holmes Jr.'s majority opinion in *Buck v. Bell*, the U.S. Supreme Court case that declared Virginia's compulsory sterilization law constitutional and created a national target that helped unite opponents. The immediate impact of *Buck v. Bell* was the crystallization of Catholic opposition to sterilization. The professions that advocated eugenics and compulsory sterilization did not entirely abandon the movement for several more decades.

The book concludes with a discussion of the ongoing attempts by American activists, legislators, and judges to employ sterilization, either surgical or chemical. The 1960s and 1970s witnessed the end of coerced sterilization in most places around the United States and the abandonment of advocacy by every professional group that had once campaigned for compulsory sterilization laws. That should have spelled the end of coerced sterilization in the United States, but the movement was resurrected at the end of the twentieth century by a new wave of anxieties about certain citizens' sexual activities and by the invention of synthetic female hormones, marketed as Depo-Provera and Norplant. Used in women, Norplant, which is implanted under the arm, provides birth control for five continuous years, while Depo-Provera is administered as a shot and prevents pregnancy for three months after injection. Men who are injected with Depo-Provera are chemically castrated, as the levels of testosterone in their bodies decline significantly. These two drugs demonstrate that, far from ending coerced sterilization in the United States, the advance of science and the development of new biotechnologies have made it even easier to coercively sterilize problematic citizens in the name of progress.

CHAPTER 1

Nipping the Problem in the Bud

The first professionals to advocate coerced sterilization as a solution to America's social ills were physicians interested in reducing the incidence of crime, or, more accurately, in reducing the number of criminals who produced children who would themselves presumably demonstrate the weaknesses they inherited from their parents. Degeneracy, transferred from parent to child through either genetic or cultural inheritance, was a concept that drew increasing study throughout the latter half of the nineteenth century. American physicians used the term *degenerate* to describe anyone who exhibited diminished mental, moral, or sexual capacities, and they believed that the sources of degeneracy were a combination of biological and environmental factors.[1] The language used to describe such people was often brutal and dehumanizing, and it reflected the subhuman status many authors ascribed to the degenerate. Decades before any American biologist ever advocated the study of eugenics, long before Francis Galton coined the word *eugenics*, and even years before Charles Darwin published his theory of evolution by natural selection, American medical doctors considered the merits of sterilizing the nation's degenerates as a method to penalize them, to prevent them from committing future crimes, to reform them, and to prevent the propagation of their kind.

The Memorial

The earliest American advocate of sexual surgery to control or eliminate social ills appears to have been Gideon Lincecum. A prominent Texas physician and ardent proponent of castration, Lincecum was decades ahead of most of his colleagues in linking the topics of animal breeding, human health, and social policy. In 1849, he authored a bill for the Texas legislature, which he called "the Memorial." The bill would have substituted castration for execution as penalty for certain crimes. Lincecum published and mailed copies of it to nearly 700 Texas politicians, journalists, prominent citizens, and medical doctors. He described how the threat of castration would serve the punitive purposes of deterrent and punishment at least as well as execution, plus it would serve as a check on the criminal type by preventing them from propagating their kind.[2]

Written ten years before Darwin's *On the Origin of Species*, Lincecum's Memorial demonstrated that even though Darwin's work was used by some to justify eugenics and coerced sterilization, evolutionary theory was not at all necessary for the development of compulsory sterilization laws in the United States. Using arguments that would become common among eugenicists half a century later, Lincecum discussed how selective breeding could improve the human race by preventing the lowest citizens of the state from reproducing. "Like begets like," he wrote to another physician. "The laws of hereditary transmission cannot be overruled. When the horse and the mare both trot, the colt seldom paces." Likewise, he reasoned, these laws were applicable to breeding better humans: "To have good, honest citizens, fair acting, truthful men and women, they must be bred right. To breed them right we must have good breeders and to procure these the knife is the only possible chance."[3]

Lincecum's Memorial was widely discussed after being published in most Texas newspapers and in the *Colorado Democrat*. Much to his dismay, the proposal was generally rejected and often made the object of jokes. The bill was introduced in the Texas state legislature in 1855 and again in 1856, and Lincecum reported that the legislators "did it in a manner better calculated to excite ridicule and opposition than a philosophical consideration of the matter." Ultimately, the Memorial was referred to the committee on stock raising. Lincecum responded to the criticism and mocking dismissal of his plan with a joke of his own: the Lincecum Law "can not progress as rapidly as it should without the aid of the press. . . . But the Press must have the benefit of the purifying implement itself before they can be moved to the advocacy of righteousness."[4]

As was the case in many other states, Texas judges, juries, medical authorities, and vigilantes did not need the legislature's approval to carry out coerced sexual surgeries. In 1864, for example, a jury in Belton, Texas, found a black man guilty of rape and sentenced him to "suffer the penalty of emasculation."[5] From the 1860s through the 1880s, newspapers reported the castration of men who were convicted of rape and of violent assaults on suspected rapists, almost all of whom were African Americans. Likewise, authorities in Kansas, Oregon, Indiana, Pennsylvania, and Illinois reported coerced sterilizations of prison inmates and mental health hospital patients without the legal umbrella of compulsory sterilization laws.

The first widely cited advocate of sexual surgeries as a solution to complex personal and social problems, and therefore an author we can justifiably identify as one of the founders of the American sterilization movement, was Orpheus Everts, the superintendent of the Cincinnati Sanitarium. In a paper presented to the Cincinnati Academy of Medicine in 1888, Everts explained how he had "given the subject of emasculation a good deal of attention in the course of twenty years' constant association with, and study of, one of the several defective classes of society." He believed that castration, while not a proper penalty for rape, was appropriate for those who were convicted of crimes that indicated "constitutional depravities that are recognized as transmissible by heredity." A law that allowed for the surgical asexualization of certain criminals would, he argued, "eventuate in an effectual diminution of crime and reformation of criminals." Everts offered a long list of claims to support his assertion that castration would deter prospective criminals, remove the power of confirmed criminals to rape again, and "diminish the number of the defective classes of society by limiting, to the extent of its application, the reproductive ability of such classes." Most of his supporting claims emphasized the basic animal nature of humans, using Darwinian language about "the struggle for existence, in which the fittest, most capable, survive; and the unfit, deficient, perish." He concluded by contrasting asexualization with the death penalty, which "goes beyond the necessities of the case or the requirements of Nature and destroys the man."[6] With execution as a widely accepted means of punishment, castration seemed a much less drastic intervention.

After Lincecum's initial campaign in favor of the social benefits of sexual surgeries, it took nearly half a century before enough American medical doctors grew interested in the subject to justify calling it a movement. Everts's paper was widely read and often cited among pioneering sterilization advocates. Take, for example, the *Cincinnati Medical News* publication of a report from the Detroit Medical and Literary Association's 1890 symposium on the prevention of conception. One participant, identified only as Dr. Stevens, "held that prevention should be used wherever there was evidence of hereditary mental incompetence," and he stated that he believed that abstinence, rather than sterilization, should be employed because "the indulgence of sexual desires should not be a pleasure nor should they pursued for pleasurable motives." Another, a Dr. Carsten, stated that "he thought all insane and criminal individuals should be castrated," and that contraception was "practiced most by those who ought especially to propagate, least by the pauper classes."[7] Coming well over a decade before any American biologist advocated eugenics or compulsory sterilization, these claims demonstrate how physicians originated the movement to coercively sterilize people they considered inferior.

In the early 1890s, the trickle of articles on coerced sterilization in American medical journals increased considerably, and medical doctors around the country became advocates of the use sexual surgeries to penalize criminals, prevent sexual crimes, treat perverts, and prevent the unfit from perpetuating their kind. In 1891, William A. Hammond, a surgeon and retired U.S. Army general, presented a paper to the New York Society for Medical Jurisprudence in which he argued for the castration, rather than execution, of criminals and offered four justifications. First, it would provide an effective deterrent to future criminals, as "man places greater value upon his generative powers than he does upon his life, and this in a great measure independently of any desire he may have for sexual intercourse." Second, a castrated criminal, unlike one who was executed, could become "a useful member of society" and could be employed in occupations in which "courage, boldness, originality, are not essential," because such traits apparently somehow originated in the testicles. Third, Hammond claimed, castration was therapeutic because it altered "the mental organization of the wrongdoer as to remove him from the category of the criminal class and certainly to prevent acts of violence in contravention of the law." Finally, without using the term *eugenics*, Hammond asserted that castrated criminals would reduce the number of future criminals. Referring to the Jukes family, he explained how "several hundred criminals would never have been born" and "many crimes would have been prevented, and the State would have saved the expenditure of a great deal of money."[8]

In Texas, calls for the sterilization of criminals that were originally offered by Lincecum were again taken up in the 1890s by Ferdinand Eugene Daniel, a battlefield surgeon and veteran of the first Battle of Bull Run. Daniel served in both the Confederate and U.S. armies before moving to Galveston, Texas, to teach as a professor of anatomy and surgery at the Texas Medical College. In 1893, he presented the paper "Castration of Sexual Perverts," which was later published in the *Texas Medical Journal*.[9] Daniel was the first American professional to bring together the critical elements of the argument in favor of compulsory sterilization: claims about the alarming rate of increase in the number of defectives, the link between sexual perversions and biological or mental inferiority, and the potential of sexual surgeries both to treat existing problems and eliminate the source of these problems from future generations. Clearly influenced by Everts's 1888 paper on the subject, Daniels described castration as both humane and effective in deterring crime. He also addressed the legal aspects of forcibly castrating mental hospital patients and prisoners by describing a conversation he had with Texas governor J. S. Hogg, who had previously been attorney general of Texas. Governor Hogg, he explained, "assured me that there is not a doubt of the legal right on the part of the superintendent of an insane asylum to castrate a patient for mental trouble, if, in his judgment, it be necessary or advisable." Many medical professionals heard Daniel's arguments throughout 1893 because, in addition to speaking before the joint session of the World's Columbian Auxiliary Congress as well as at the International Medical-Legal Congress and the New York Chapter of the Medical-Legal Society, his article was printed in the *Texas Medical Journal*, the *Medico-Legal Journal*, and the *Psychological Bulletin*.[10] The discussion that followed Daniel's presentation at the World's Columbian Auxiliary Congress demonstrated that there was near unanimous support among his colleagues in favor of compulsory castration. His presentations to professional audiences, the appearance of his paper in numerous journals, and the other authors on the subject he cited in his work showed that by the early 1890s, more than a decade before American biologists began advocating eugenic sterilization, American physicians were ardently campaigning for it.

A year after Daniels's work appeared, A. C. Ames of Plymouth, Nebraska, published "A Plea for Castration, as a Punishment for Crime" in the *Omaha Clinic.* In it, he asserted that "there is as much evidence to prove that vice and crime are hereditary as there is to prove any disease to be so," and "society has a right to protect itself against crime." The solutions for the nation's problems with crime were limited to "castration, execution or imprisonment for life of criminals." Of the three, he argued that castration was the least severe. He also made clear that he was "in favor of extending the practice to women, for the same degree of crime as in men, and removing the ovaries, if the woman was to be set at liberty at less than 40 or 45 years of age."[11] That same year, F. L. Sim, a medical doctor in Memphis, Tennessee, published "Asexualization for the Prevention of Crime and the Arrest of the Propagation of Criminals" in his state's medical journal. Just as Ames contrasted the brutality of execution with what he identified as the more humane act of castration, Sims began his argument in favor of castration by comparing it to capital punishment. Castration would cure men of mental neuroses, adequately punish rapists and sodomites, and discourage mentally or morally weak men from committing sexual crimes. Castration, he concluded, "is the proper method for the immediate and permanent protection of society, the punishment of criminals, and the arrest of their propagation."[12] Like most of the nineteenth-century arguments in favor of coerced sterilization, Ames's claims relied on anecdotal evidence and commonsense notions that "like begets like" to justify limiting some people's ability to reproduce.

A year later, in 1895, B. A. Arbogast, a medical doctor from Breckenridge, Colorado, published "Castration the Remedy for Crime" in his state's medical journal. Arbogast argued that the "only worthy aim of a system of relief is the restoration to the ranks of normal . . . [of] those criminals who are capable of such restoration, and the speedy extinction of those who are beyond the possibility of such help." Physicians, he claimed, were the proper social authorities to suggest remedies to vice and crime and to apply the treatment. No right-thinking American could object, Arbogast concluded, to the castration of "the psychosexual monster, the Sadist, and the rapist," because "once his lust is aroused it is entirely beyond his control and the greatest cruelty is resorted to accomplish his hellish design." Ultimately, he believed that American liberties, the high standards of American citizenship, universal justice, and the evolution of a nobler humanity would all be served by castrating criminals.[13] That same year another physician, E. Stuver, of Rawlins, Wyoming, published an article in the *Transactions of the Colorado Medical Society* on the use of castration to limit disease and to both prevent and punish crime. Believing that "many of the most common and revolting crimes are directly traceable to sexual perversions," Stuver concluded that castrating criminals would serve as a powerful deterrent as well as remove or modify "their exciting causes or favoring circumstances." Discussions that followed his presentation at the annual meeting of the Colorado Medical Society demonstrated that while there was some contention about whether castration could actually reduce the incidence of crime among the castrated, there was general agreement that it would benefit society because there would thereafter "be no further propagation of his kind."[14] Finally, A. C. Corr's 1895 "Emasculation and Ovariotomy as a Penalty for Crime and as a Reformatory Agency" summarized a symposium held a year earlier by the Illinois State Society to discuss various methods for punishing, reforming, and executing criminals. Last on their list of six methods of dealing with crime was castration: "In unsexing all constitutionally depraved convicts, the most important of all results is reached, the limitation of the productive capabilities of this class, thus aiding natural selection, and insuring if extensively applied the survival of the fittest." Corr concluded by suggesting that castration be used not just to punish rapists but also those who commit "the most heinous of crimes," seduction, which he described as premeditated rape.[15]

Sexual surgeries to reduce crime and to treat both mental illnesses and sexual perversions also emerged as a central tenet in a late-nineteenth-century medical movement called orificial surgery. Popular from the early 1890s through World War I, orificial surgery was championed by Edwin Hartley Pratt, an Illinois homeopathic general practitioner. Pratt believed that "a vigorous sympathetic nervous system means health and long life." He situated a variety of disorders in or around his patients' orifices, arguing that "the weakness and the power of the sympathetic nerve lies at the orifices of the body. Surgery must keep these orifices properly smoothed and dilated." Pratt's biographer concluded, "he rarely saw an orifice that was not in need of a surgeon's scalpel" to relieve the patient of "constipation, dysmenorrheal, eczema, insanity, insomnia, tuberculosis, and vomiting."[16] After publishing *Orificial Surgery and Its Application to the Treatment of Chronic Diseases* in 1887, Pratt organized the American Association of Orificial Surgeons, which held annual meetings until 1920; he invented and sold surgical instruments for orificial surgery; and he founded and edited the *Journal of Orificial Surgery* from 1892 through 1901.[17] In the first volume of the journal, Pratt presented the basic beliefs on which he founded the philosophy of orificial surgery. Contrasting his view of the human body with the simple anatomy studied by most medical doctors, Pratt described the human form as an "intricate and delicate interweaving of several human forms, the blending together of which constitutes the individual which is to be the object our study." Pratt believed that there were two nervous systems, the cerebral-spinal and the sympathetic, and that the best treatment for disease balanced the two. Orifices were critical locations because "food, drink and air supply our hopes for the future. The alvine canal, urinary tracts, the sweat glands and our expirations relieve us of what has been." Maintaining the orifices was therefore critical to health, and Pratt advocated surgery to smooth or open constricted orifices.[18] Pratt and other orificial surgeons believed that their approach was a valuable tool in treating "cases of perversion of sexual instincts." In the first volume of his journal, Pratt explained that "in a majority of cases all spontaneous and unnatural sexual activity can be removed by securing by means of a little judicious pruning an ideal condition of the lower openings of the body." In the same volume he wrote, "Some form of rectal or sexual irritation is at the basis of all cases of perversion of sexual instincts. It is no more natural to be lustful than it is to limp."[19] The orificial surgery movement peaked around the turn of the century and faded as its aging constituency was unable to attract new members in the years between 1910 and 1920.[20]

By the mid-1890s, American medical doctors had established a large and growing movement in favor of castrating a class of citizens they believed were responsible for most of the nation's crime and, they argued, would produce the majority of the next generation's criminals. Their targets, broadly described, were defectives, those who lacked the moral or intellectual capacity to control themselves and act appropriately civilized. Ten years after physicians threw their weight behind a movement to legalize the sterilization of the nation's defectives, American biologists met with plant and animal breeders in St. Louis to establish the American Breeders Association, the first professional organization that included a committee specifically designed to advance the study and public understanding of eugenics. For American medical doctors, the social value of sterilization was by then old news.

Figure 1.1. Edwin Hartley Pratt (1849–1930). Taken from the frontispiece of his *Ortificial Surgery*.

Targeting Women

Women as well as men were targeted for sexual surgeries to relieve them and society of a variety of burdens that their medical providers identified. In 1886, Dr. S. C. Gordon, chairman of the Section of Obstetrics and Diseases of Women and Children of the American Medical Association, presented "Hysteria and Its Relation to Diseases of the Uterine Appendages" at the organization's annual meeting. Citing claims from both modern and ancient authorities, Gordon offered Hammond's assertion from *Diseases of the Nervous System* that the "phenomena of hysteria may be manifested as regards the mind, sensibility, motility and visceral action, separately or in any possible combination." After presenting several case studies to demonstrate his claim that most women's hysteria was founded in diseased organs of the reproductive system, Gordon explained that he had performed twenty-five hysterectomies, most as treatment for hysteria. In summarizing his beliefs about the potential of sexual surgery to aid the mental condition of women, he concluded that "hysterical symptoms occur almost exclusively in women," so one could presume that these symptoms were "due to disease of some organ or organs peculiar to women." He hoped that "the old idea that a hysterical woman is only to be laughed at, and treated as one who deserves no consideration at our hands," would soon fade and that women would be properly treated by medical professionals who recognized the actual cause of their hysteria.[21]

In 1886, the same year that Gordon published his article, the *American Journal of Medical Sciences* published three papers presented at a symposium titled "Castration in Mental and Nervous Diseases," all of which focused on the castration of women via ovariotomy and oöphorectomy. The papers, by an Englishman, a German, and an American, demonstrated the fact that while late-nineteenth-century Europeans showed little interested in sexual surgeries for resolving mental and nervous disorders in women, at least some American medical providers were enthusiastic about their potential. The first of the three papers, presented by T. Spencer Wells, the former president of England's Royal College of Surgeons, discussed the history of the operations from the 1870s through the early 1890s and criticized their use for the treatment of mental or emotional disorders. With a mortality rate of nearly 15 percent, half of which were due to septicemia, and of no apparent use in the treatment of nervous diseases, Wells concluded, "That in nearly all cases of nervous excitement and madness," hysterectomies were "inadmissible."[22] Alfred Hegar, professor of obstetrics and gynecology at the University of Freiberg, similarly criticized the use of sexual surgeries to alleviate nervous disorders in women. He accepted that "a group of symptoms in the nerves of the genitals and in their vicinity" could account for nervous disorders and that "such a condition during the exploration may be overlooked, but who would undertake an operation fraught with danger to life on such possibility?"[23] In sharp contrast to his European counterparts was Robert Battey, an American physician who had trained in Philadelphia and practiced in Georgia. Battey served as a surgeon in the Confederate army and as professor of obstetrics the Atlanta Medical College, and he edited the *Atlantic Medical and Surgical Journal.* He was a pioneer in operating on the urinary organs of men and women, and in his contribution to the symposium, Battey described how he had castrated women for any one of three classes of mental and nervous disorders: oöphoromania, oöphoro-epilepsy, and oöphoralgia. Each of these disorders, he explained, was caused by "nervous irritation proceeding from the ovaries." Battey tabulated the results of the thirty-six women whose ovaries he had removed, and he determined that nearly two-thirds of them had been cured by the operation, while less than 17 percent showed no improvement whatsoever. He concluded with a series of case studies to demonstrate the usefulness of castration in treating mental and nervous diseases in women.[24]

Hysteria was only one of several supposed ailments that nineteenth-century medical doctors treated with sexual surgeries. Another was sapphism, a term used by medical professionals in reference to homosexuality among women. In this context, the professional literature often discussed the case of Alice Mitchell, a woman in Memphis who had been accused of killing another woman, Freda Ward. An investigation uncovered the fact that there had been "an unnatural affection existing between Alice Mitchell and Freda Ward," and Mitchell "seems to have been the ardent one." They had arranged to be married, with Mitchell costuming herself as man, but Ward's friends intervened and convinced to her cancel the ceremony. In response, in July 1892 Mitchell attacked Ward, cutting her throat with a razor.[25] Evaluations by medical doctors, psychiatrists, and journalists emphasized the fact that Mitchell's "sexual abominations" as well as her ability to murder her former lover demonstrated how badly defective she was. She became for some an example of the sort of woman best suited for asexualization.[26]

In addition to performing hysterectomies to relieve hysteria or prescribing sexual surgeries for sapphism, the alleged sin of masturbation motivated some of the operations on women. C. A. Kirkley, a medical doctor from Toledo, Ohio, published "Gynecological Observations in the Insane" in the *Journal of the American Medical Association* (*JAMA*) in 1892. Kirkley observed the cases of nearly 600 women at the Toledo Hospital for Insane in 1890 and 1891 to determine how diseases of their reproductive organs and removal of those organs might "effect relief upon the mental condition" of the insane women. "That insanity exercises a peculiar influence upon the sexual organs of women," he claimed, "there can be no doubt. This may also be said of insanity in men." In an interview with an attendant at the hospital, he learned that nearly 39 percent of the women in the institution "admitted to her that they practiced masturbation whenever the opportunity presented itself." This "unnatural habit," he explained, was generally only practiced by older women, as "it is extremely doubtful if any pure-minded young girl has the slightest idea of sexual desire previous to her marriage." Ultimately, Kirkley recommended increased study of the role of the sexual system in causing or helping cure insanity in women, and he concluded, "Every insane hospital should have its gynecologist."[27]

The Crime against Nature

While some of the professional literature addressed the use of sexual surgeries to solve the vexing problems of homosexuality and masturbation in female patients, these problems caused considerably greater anxiety among medical providers when their patients were male. From the origins of the American sterilization movement through the late 1920s, men were disproportionately targeted for sexual surgeries. While women who masturbated or practiced sapphism were singled out for special discussion in the professional literature, men who acted similarly were treated as particularly dangerous to civil society. Men diagnosed as "chronic masturbators" were charged with committing the sin of onanism, a term derived from the biblical story in Genesis. Onan's brother, Er, had been "wicked in the Lord's sight; so the Lord put him to death." Onan was then ordered to impregnate his dead brother's wife, Tamar. Not wanting to father children with Tamar, whenever Onan "lay with his brother's wife, he spilled his semen on the ground" instead of providing children for his brother's widow. "What he did was wicked in the Lord's sight; so he put him to death also."[28] In medical professionals' literature from the late nineteenth century, the sin of onanism described the act of wasting one's seed; that is, masturbating.

The so-called crime against nature was a collection of activities, all sexual in nature, committed by allegedly deviant men that were so universally despicable that a euphemism was necessary even in the actual language of the laws. A man could potentially engage in the crime against nature with another man, with a woman, or with animals. The term generally referred to anal sex and was founded on sixteenth-century English prohibitions against buggery, a crime "not to be named among Christians" and punished with "death without the benefit of clergy." The crime against nature also included the act of sodomy, which usually included anal sex and sometimes referred to oral sex as well, again with a man, woman, or animal. Exact definitions of these terms varied considerably from state to state. For example, in a 1904 case, the supreme court of Georgia defined sodomy as "carnal knowledge and connection against the order of nature by man with man, or in the same unnatural manner with woman" and included both anal and oral sex.[29] Ten years later, the supreme court of North Carolina declared that sodomy referred to anal, but not oral, sex. However, in the same case the court declared that the crime against nature included both acts.[30] In 1953, a legal scholar concluded from an analysis of court decisions that North Carolina's statute outlawing the crime against nature included bestiality, anal sex, oral sex between two men, and perhaps oral sex between a man and a woman or two women.[31] Whatever the precise definition, these acts were vilified throughout the professional literature, and sexual surgeries were often offered as their appropriate cure.

From the 1880s through the early twentieth century, some medical doctors concluded that castration was the best treatment for men compelled to commit sodomy or the crime against nature, however either was defined. For example, in 1914 Charles H. Hughes described how he had operated on a "gentleman of ordinary moral, intellectual and physical parts and psychic impulsions, save for the affliction which distinguished him"; namely, he had "reciprocal homo-sexual associates." The patient twice requested that Hughes operate on him, the first time to excise the penis nerve and the second time to castrate him by excising the testes.[32] Throughout the late nineteenth and early twentieth centuries, men were disproportionately targeted for coerced sterilization, especially those men who had been suspected or convicted of committing any number of sex crimes, ranging from masturbation to rape. In the late 1920s, however, the number of women surpassed the number of men coercively sterilized in the United States each year, and by midcentury, sexual surgeries were rarely justified simply by homosexual activities or masturbation. Both acts were commonly mentioned as part of a collection of distasteful activities of which mentally ill or mentally deficient patients were accused.

Throughout the twentieth century, American social scientists as well as legal and medical professionals occasionally discussed the use of chemical or surgical castrations to treat homosexuality among males and to treat, punish, or control sexual criminals, especially rapists and child molesters. For example, officials in both Oregon and North Carolina ordered castrations for men convicted of sodomy, child molestation, and rape into the 1940s.[33] In 1927, R. L. Steiner, superintendent of the Oregon State Hospital, stated at an annual meeting of the Oregon State Medical Society, "Nothing less than castration or complete unsexing will do the rapist any good. The same applies to the sexual pervert or the chronic masturbator."[34] Some social scientists continued to advocate castration into the latter half of the twentieth century. In 1953, Karl Bowman and Bernice Engle published "The Problem of Homosexuality" in the *Journal of Social Hygiene.* They argued that multiple factors caused homosexuality, including genetics and certain aspects of personality development. They listed potential treatments, such as hormone therapy, metrazol convulsions, and electroconvulsive shock therapy, but most strongly emphasized the use of castration as was done in Scandinavian countries, Switzerland, the Netherlands, and the United States. "Therapeutic castration," they concluded, "therefore seems to be a valid subject for research, under carefully controlled scientific study."[35]

The First Opposition to Coerced Sterilization

Discussions about the use of sexual surgeries on both men and women to solve various ailments were common enough among American physicians in the 1880s and 1890s to warrant some criticism by opponents. For example, the *JAMA* published a critical evaluation of efforts at the State Hospital for the Insane in Norristown, Pennsylvania, to solve women's mental problems with oöphorectomies.

Dr. Joseph Price and Dr. Alice Bennett, with the consent of the hospital's trustees, reported that they had, on October 29, 1892, removed the ovaries of four women and had "marked for operation" fifty other wards. In response, the Committee on Lunacy of Pennsylvania State Board of Public Charities published a report on the question, explaining that in both the United States and Europe operations to remove ovaries had been performed "in the hope that the mental and physical diseases would be cured." The *JAMA* editors explained, "The increasing frequency of these experimental mutilations and their doubtful ultimate success, have caused conservative medical opinion to halt and to dispassionately discuss the whole subject." They doubted that the relative or guardian of an insane women had either the moral or the legal right "to give consent to the unsexing of the insane person, whose power to give or withhold consent is temporarily or permanently in abeyance." The editors concluded that "insanity is not a disease of the ovaries, nor of any other part of the body which is accessible to the surgeon's knife."[36]

Very few individual physicians spoke out against the use of sexual surgeries to solve complex psychological, sexual, and social problems in the nineteenth century. One of the few was Mark Millikin, whose 1894 article for the *Cincinnati Lancet-Clinic* condemned castration as mutilation. Summarizing articles from five different medical journals, each of which advocated castration as a treatment for criminals and sexual perverts, Millikin concluded that members of "castration coterie" overlooked the role of the environment in producing criminality and perversion. "Negroes," he claimed were "the typical product of a bad social system," and one ought not expect that "the habits derived from savagery and bondage" would be "changed by thirty years of freedom." Sexual perverts and white criminals were caused by "over-crowded tenements, rum, and the denial of the natural opportunities to which all men have rights."[37]

The most widely criticized use of sexual surgery in the late nineteenth century focused on the work of Dr. F. Hoyt Pilcher, a physician and superintendent of the Kansas State Asylum for Idiotic and Imbecilic Youth in Winfield, Kansas. Prominent in Populist politics and trained as a clinician, Pilcher took the position as superintendent in 1893 and introduced medical, rather than educational or behavioral, remedies for severely mentally disabled patients by using castration as a treatment for what he referred to as self-abuse. Authorities at the asylum had long considered masturbation a serious problem among their patients, and an earlier superintendent, C. K. Wiles, had dealt with what he termed "a nameless habit" by prescribing constant supervision for all chronic masturbators. Even that was not enough to stop one patient from masturbating, so Wiles improvised a straitjacket, which consisted of a canvas bag with arms that were buckled together during the night.[38] After consulting with other physicians, Wiles's successor, Pilcher, castrated chronic masturbators to relieve them of the apparently unmanageable burden of trying to restrain themselves.[39] Between 1893 and 1898, he amputated the testicles of forty-four males and performed hysterectomies on fourteen females, whose average age at the time of the operation was twenty years.[40]

Pilcher's prescription for chronic masturbation was not out of line with the professional discourse of his time and place. Throughout the 1890s, both before and after he began performing sexual surgeries on inmates of his institution, other Kansas medical doctors had advocated the castration of criminals and defectives. In 1890, R. E. M'Vey, professor of clinical medicine at the Kansas Medical College, read a paper before the Kansas Medical Society titled "Crime—Its Physiology and Pathogenesis. How Can Medical Men Aid in Its Prevention." After describing the various sources of criminals, such as murderers, who "are a class of men with a low development of the intellectual and moral facilities," and rapists, whose crime "grows out of an unbalanced condition of the mental and reproductive functions," M'Vey offered a two-pronged remedy: improve both the hereditary basis that is the root source of crime and eliminate environments that nurture the criminal instinct. Ultimately, he concluded, "the proper plan for the prevention of crime is the commencement, a hundred years before the criminal is born, to better this heritage and environment."[41] Seven years later, another Kansas medical doctor, Bernard Douglass Eastman, superintendent of the Topeka State Insane Asylum, read a paper at the annual meeting of the same state medical society titled "Can Society Successfully Organize to Prevent Over-Production of Defectives and Criminals?" Like M'Vey before him, Eastman emphasized the role of heredity in forming the foundation of mental and moral attributes, asserting that they are "derived from progenitors and ancestors with more or less variation in intensity." He concluded that despite practical difficulties that could be overcome "by a slow process of education in which scientific motherhood, untainted fatherhood, the same code of morals for both sexes, the universal acceptance of the golden rule, and utter subordination of the individual to the wellfare [*sic*] of the public," medical doctors ought to begin a campaign of "asexualization of criminals and defectives."[42]

Despite support from many of his colleagues, Pilcher's sexual surgeries on patients at the Kansas State Asylum created significant protest from journalists and the public in his home state. The *Winfield Daily Courier* devoted the entire front page of its August 24, 1894, issue to railing against Pilcher, claiming in its headline, "Mutilation by the Wholesale Practiced at the Asylum" and promising "Names and Full Particulars." The editors did indeed provide the names and hometowns of eight young men castrated at the Kansas State Asylum and explained that four others had been castrated, but they were unable to obtain their names. After a short history of the institution, which was lauded for its charitable work, Pilcher came under direct attack. It is obvious that even while it claimed that the article was "not written for political effect," the newspaper's editor was ardently opposed to populism and linked Pilcher's politics with his personal and professional shortcomings. The extended article claimed that Pilcher was "addicted to the use of liquor" and "intemperate in his habits" and was "unworthy to fill a position of such high trust and responsibility as that involved in the care of these poor, helpless, idiotic children." The author openly accused Pilcher of raping the young women in his charge and, in addition to providing eyewitness testimony from doctors who assisted Pilcher in performing eleven castrations, listed inmates who had died in Pilcher's charge.[43]

While the public, journalists included, objected to Pilcher's work, medical authorities in Kansas as well as in other states were generally supportive of his attempts to "medicalize" the treatment of mental health patients who had not benefited from more modest therapies. In 1894, the *Kansas Medical Journal* published a discussion of Picher's work along with a description of the public uproar over his decision to castrate those boys he believed were addicted to masturbation. It discarded the claims from the *Topeka Lance* that the castrations were politically motivated to "envenom the voter against the administration under whose reign these operations were accomplished." In sharp contrast to the journalists' attacks on Pilcher, the editors of the *Kansas Medical Journal* called him "a brave and capable man" who could give his charges "a restored mind and robust health."[44] Throughout 1894, Texas medical doctors likewise discussed the efficacy and morality of Pilcher's treatment. The editors of the *Texas Medical Journal* reprinted the procastration editorial from the *Kansas Medical Journal* as well as a "Dear Family Doctor" column from the *Kansas Farmer* that supported the use of castration as a treatment, as a deterrent, and for its eugenic potential.[45] They introduced both pieces with an explanation that Pilcher had castrated "a number of boys confirmed in the evil habit of masturbating" and had been castigated in the local press for performing operations considered cruel, brutal, and unjustifiable. An overview of the material, the editors claimed, demonstrated that "there is a growing sentiment in the profession in favor of castration," and that "the public are not prepared for anything of the kind and must be educated up to it." They hoped the uproar would give the public a chance to hear from experts about the value of castration in improving society.[46]

Several years after Pilcher stopped asexualizing boys and girls at the Winfield asylum, F. C. Cave produced a short description of the long-term effects of the surgeries. Of the forty-four boys castrated at the institution, half of them remained patients there, while all fourteen of the girls who had received hysterectomies were still residents. Cave explained that he considered procreation inadvisable for every one of the inmates under his care; nonetheless, over half of them were still capable of producing children, which suggests that the operations continued after Pilcher left the institution. Of the three dozen who had been asexualized, he saw no special change in their mental conditions, but he identified significant moral improvement because "they are not addicted to onanism and other prevalent perversities." This was not because "their standard of morality has been elevated," but because "the elimination of the physical factors has caused the betterment." In women, the operation caused menstruation to cease and breasts to atrophy, and eliminated their desire for sexual intercourse. Several of the women had epilepsy, and the operations appeared to have no impact on their ailment. Among the men, one in particular, feminine qualities of fair skin, higher pitched voices, and changes in body contour appeared, and "all sexual desires have been lost and they are impotent in every sense of the word." Cave advocated the continued use of asexualization surgeries, preferring oöphorectomies for women and testiectomies for men. He did not believe that vasectomies were of any value, as "the act of copulation is not prevented and it seems to me this fact would tend to increase sexual debaucheries as the pleasure would not be lessened and the danger from conception would be eliminated." Cave concluded by arguing that there was little need to sterilize degenerates who would never be released from public institutions. However, he argued that "the delinquent who is not confined is the individual who needs surgical treatment," and he favored a compulsory sterilization law because "it is time some drastic action were taken to stem the ever increasing tide of weak-minded individuals who are demanding more and more room in our charitable institutions by their increase."[47]

Histories of the American eugenics movement and of compulsory sterilization in the United States strongly emphasize the role of the development of the vasectomy in 1897 and the rediscovery of Mendel's work three years later. The claims made by American physicians throughout the latter half of the nineteenth century demonstrate that a significant number of health care providers advocated sexual surgeries for both therapeutic and eugenic reasons well before the turn of the century. These medical doctors saw in the coerced sterilization and asexualization of their patients remedies for individual and social ills, and they aggressively campaigned for the legal authority to perform the operations.

Racism and Castration

Interest among American medical doctors about the potential use of castration to solve complicated social and medical problems was not an isolated activity or something limited to trivial figures, nor was it free of overt racism. In 1893, the president of the American Medical Association, Hunter McGuire, wrote to G. Frank Lydston, professor of genito-urinary surgery and syphilology at the Chicago College of Physicians and Surgeons, asking him to provide "some scientific explanation of the sexual perversion in the negro of the present day." After reading the chapter on perversion in Lydston's 1892 *Addresses and Essays* and motivated by the increasingly common newspaper reports of "the crime of a negro assaulting a white woman or female child," McGuire sought Lydston's advice on the best possible solution to the problem and lamented the "innocent, mutilated, and ruined female victim and her people."[48] McGuire's original query and Lydston's reply were printed in book form in 1893 under the title *Sexual Crimes among the Southern Negroes*, which nicely demonstrated McGuire's zealous racism and Lydston's appreciation for castration as a solution to complex and vexing social problems. In contrast to McGuire's racist claims about how "the negro is deteriorating morally and physically," Lydston argued that the supposed extraordinary perversion of African Americans "cannot, in the strictest sense of the term, be justified scientifically." Instead, he argued that the average "Negro compares quite favorably as regards sexual impulses—taking all abnormalities into consideration—with the white race." In the context of rape, therefore, Lydston treated blacks and whites equally, believing that each race had its fair share of degenerates. The best possible treatment, he concluded, was not execution because he "failed to see wherein capital punishment has, in the aggregate, repressed those crimes for which it is prescribed." Instead, Lydston believed that there was "only one logical method of dealing with capital crimes and criminals of the habitual class—namely, *castration*." Castration would prevent "the criminal from perpetuating his kind," especially when the operation was "supplemented by penile mutilation according to the Oriental method." Moreover, castrated felons would "be a constant warning and ever-present admonition to others of their race."[49]

Lydston's advocacy of castration continued well into the twentieth century. Published in 1904, *The Diseases of Society* devoted forty pages to a discussion of the "principles of evolution in their relations to criminal sociology and anthropology, and to social diseases in general."[50] Referencing Darwin's work, Lydston argued that degenerate humans were similar to the lower animals in that they lacked the ability to overcome instinctual urges. The emergence of morals, social norms, and ultimately theology, religion, and law marked humanity's ascent out of the animal world. In using the term *man*, Lydston meant just that; with respect to morals and to crime, a woman, he claimed, "resembles the child in her emotional instability, but her will is relatively weak, so that she is often very like the child in her disregard of property rights, selfishness, and utter lack of altruism."[51] Vital to evolution was heredity, and likewise heredity was vital to understanding the evolution of the criminal. Stressing the notion of criminal evolution, Lydston argued that if heredity does not hold good for the production of criminals, "then it fails everywhere else."[52] One chapter of the book was devoted to the "therapeutics of social disease" and discussed the treatment of crime via two general methods: control of heredity through marriage control, and asexualization or sterilization and the general improvement of environmental conditions by the passage of laws to help the poor, management and reform of juvenile delinquents, increased education, and careful selection of punitive measures. Describing the marriage license window as the seat of "self-contamination" and the "fountain-head of the stream of degeneracy that sweeps through all social systems," he claimed that the "marriage license is the agent that sets the individual and social machinery for the manufacture of degenerates in operation."[53] So-called sanitary marriages are the ideal, Lydston argued, but are rare. Society had the right to "defend itself against the finished product of its matrimonial factory of degenerates" by restricting by law those with sexually transmitted diseases, histories of drunkenness, and epilepsy from marrying unless "they submit themselves to sterilization."[54] Even though the vasectomy had by then been widely publicized for several years, Lydston still advocated castration. He accepted that "absurd sentimental objections" against castration were strong and recognized that "the same results in the prevention of degeneracy can be obtained by" vasectomy and tubal ligation, but he dispensed with any sentimentalism toward the criminal by returning, again and again, to the comparison of an operation under anesthesia performed by competent physicians with "the average execution, and more especially with bungling execution."[55]

By the end of the nineteenth century, the decades-old conversation among medical authorities about the use of sexual surgeries for punishment, treatment, or prevention of certain crimes moved into the legal profession. In 1899, Daniel Brower of Chicago read a paper at the American Medical Association titled "The Medical Aspects of Crime," in which he argued in favor of the use of vasectomy for eugenical purposes.[56] That same year, an anonymous article in the *Yale Law Journal* discussed the use of whipping and of castration as punishments for crimes. Its author argued that both practices were ancient and had their place in contemporary law enforcement. Castration, the author asserted, was an acceptable punishment for rape as well as for the "certain crime of which one seldom speaks," which could have been child molestation, homosexual activities, or masturbation. Identifying the racial tensions that were often at the root of the issue, the author explained that rape was "a daily terror" to every woman in the South, and was "the cause of most of those lynching cases which disgrace our civilization." The author dismissed the two principal objections—that castration was cruel and that its use effectively lowered a human life beyond recovery—and argued that it was no more cruel when performed on a man than when it was done on cattle. "It is an adjustment to their environment in society. It is necessary to make it safe to keep them about us." Without using the word *eugenic*, the author argued that castration was "a possible and permissible mode of preventing the propagation of a degenerate class of imbeciles or paupers." Moreover, castrations were "actually done in a quiet way by not a few of the medical profession" in public institutions to prevent their charges from falling victim to their own disorders as well as to "end the line of a family which is misusing the earth."[57]

The mingling of punitive and preventative motivations with potential therapies was obvious in the claims made by a number of physicians who advocated the sterilization of criminals. For example, Jesse Ewell, a medical doctor from Ruckersville, Virginia, read a paper before the annual meeting of the Medical Society of Virginia in 1906 in which he called for castration of black men who sexually assaulted white women. Forty years of "enforced citizenship and free education have utterly failed to better the condition of the negro. He has retrograded physically, morally and mentally, which proves that there is something wrong with the system under which he lives." Contrasting castration with lynching, Ewell claimed that medical doctors, who were "conservators of the public weal," should advocate the castration and the cutting off of the ears of black men who assaulted white women. Instead of allowing citizens to shoot or hang black men accused of rape, Ewell called for the Virginia legislature to "protect our loved ones" by empowering medical doctors to castrate, mutilate, and stamp the black rapist with " 'the mark of Cain' on his forehead," which would "deter more would-be criminals than would the shooting or hanging of five hundred."[58]

Enter the Vasectomy

Harry C. Sharp is widely, but incorrectly, regarded as the originator of the vasectomy.[59] While he was among the first American advocates of the eugenic value of the vasectomy and probably the first to actually perform the operation with eugenic motivations, it is not at all clear who first attempted the operation. The vas deferens, the object of ligation in a vasectomy, was named by Berengarius of Carpi in the sixteenth century, but it was not until 1830 that Astley Cooper presented a complete study of the structure in his *Observations on the Structure and Diseases of the Testis*.[60] A. J. Ochsner, chief surgeon at St. Mary's Hospital and Augustana Hospital in Chicago, reported that in 1897 he had intentionally severed the vas deferens of two patients in connection with prostate operations. Seeing that the two patients were rendered sterile while noting no change whatsoever in their sex lives, Ochsner considered the eugenic possibilities of the vasectomy. Two years later, he published a paper titled "Surgical Treatment of Habitual Criminals" in the *JAMA*, explaining that the operation would protect the community at large, and he suggested the operation for criminals, chronic inebriates, imbeciles, perverts, and paupers. Prevented from having children, he wrote, "there would soon be a very marked decrease in this class, and naturally, also a consequent decrease in the number of criminals from contact."[61] Ochsner had effectively created a way to separate the reproductive abilities of so-called defectives from their sexual activities, and with that, the advocacy of compulsory sterilization laws as part of the American eugenics movement was possible.

In his *JAMA* article, Ochsner provided two case studies to demonstrate the need for a "reasonable plan for the surgical treatment of habitual criminals of the male sex." The cases involved middle-aged men with severely enlarged prostates, and in both cases Ochsner resectioned the vasa deferentia through one-inch incisions. The patients healed quickly and were relieved of their prostate problems, and each "found no impairment of his sexual desire or power." In retrospect, there is no reason why the vasectomy would have improved the patients' enlarged prostates. According to Philip Reilly, a medical doctor, lawyer, and author of *The Surgical Solution*, "there is no physiologic basis for supposing that severing the vas deferens should ameliorate prostatic hypertrophy, cure prostatitis, or increase sexual vigor. One is forced to attribute the improvements in the patients to chance or to a placebo effect from the surgery."[62]

Even though both of the men in Ochsner's case studies were employed, responsible, and obviously not among the targets of his colleagues' enthusiastic calls for castration, Ochsner used their cases to introduce the principal claim in his article: severing the vas deferens of "criminals, degenerates and perverts" would "do away with hereditary criminals from the father's side," it would leave the criminal "in his normal condition" aside from sterility, it would "protect the community without harming the criminal," and it could "reasonably be suggested for chronic inebriates, imbeciles, perverts, and paupers." Judging from

Ochsner's argument, he was primarily motivated in advocating the use of vasectomy to reduce the number of criminals by the writings of Eugen Bleuler and Cesare Lombroso. Unlike eugenicists who focused primarily on tainted heredity and viewed the environment merely as the context within which inherited tendencies were encouraged or repressed, Ochsner clearly believed that criminals' "surroundings must necessarily be favorable to the development of vicious tendencies." Reducing the number of children raised by criminals, regardless of any potential hereditary inclination toward crime, would reduce the overall number of criminals. It was really only necessary, he asserted, to sterilize male criminals, as female criminals generally suffer from "endometritis and salpingitis usually resulting in an occlusion of the Fallopian tubes early in their career." In cases of fertile female degenerates, he stated that "ligation and section of the Fallopian tubes has been suggested." Therefore, he concluded, it was in the best interest of "the general welfare and safety of the community, to reduce as much as possible the number of children born in the families of criminals."[63]

Sharp, a surgeon at the Indiana Reformatory in Jeffersonville, Indiana, read Ochsner's *JAMA* article with interest and shortly thereafter counseled an underage inmate named Clawson on remedies for his "excessive masturbation." According to Sharp in an interview conducted thirty-five years after the event, the boy had requested that he be castrated, but Sharp "did not feel justified in performing that mutilation," so he recommended a vasectomy to relieve him of his urge to masturbate. Why Sharp believed that a vasectomy would reduce the boy's tendency to masturbate is unclear. He reported that several weeks after the initial operation, the boy reported no reduction in his urges, so "I gave him another 'treatment' and told him to wait for six months; then, if he still desires it, I should castrate him." The boy later reported that he had stopped masturbating. The surgeon concluded that the operation somehow affected the nervous system, calming the patient and allowing him to regain self-control. "Other inmates," Sharp explained, "began to request that they have the advantage of the same operation."[64]

In the fall of 1901, Sharp presented his first paper on the use of vasectomy for therapeutic or eugenic purposes at the Mississippi Valley Medical Association meeting.[65] A few months later, he published an article in the *New York Medical Journal* on the relationship between sexual surgery and mental deficiency titled "The Severing of the Vasa Deferentia and Its Relation to the Neuropsychopathic Constitution." Crediting Francis Galton and Théodule Ribot for establishing "the fact that a general law of heredity obtains in the mental as well as in the physical life," Sharp claimed that a weak hereditary constitution was the source of a range of ailments, including hysteria, alcoholism, criminality, and insanity. Clearly influenced by Galton's *Hereditary Genius*, he contrasted hereditary defectives with families that produced "perfectly healthy" minds and explained how the rapid increase in inferior individuals was something limited to humans and unknown among animal species. In nature, "the weaklings that are unable to weather the storm" die, while domesticated animals are adequately culled by breeders, leaving "none but the perfectly healthy . . . to reproduce their kind." Judging from his 1902 article, the work of Ochsner and Brower in Illinois as well as Pilcher in Kansas motivated Sharp to begin sterilizing some of his wards at the Jeffersonville Reformatory in Indiana.[66]

Sharp's decision to sterilize dozens of young men laid the foundation for the first successful compulsory sterilization law in the United States. Following failed attempts in Michigan and Pennsylvania, Indiana enacted the nation's first sterilization law in 1907. Two years later Sharp read "Vasectomy as a Means of Preventing Procreation in Defectives" at the American Medical Association meeting in Atlanta, then published an article with the same title in the *JAMA*.[67] He argued that the public "is rapidly coming to realize that our public dependents are largely recruited from the defective classes" and that a predisposition to insanity is "an inherited defect." The cost of these defectives and their offspring, he claimed, had more than doubled between 1890 and 1908, as hereditary defects had rapidly increased in number. Castration was a "means that has been suggested for the purpose of preventing procreation in the unfit," but it was "of too much gravity and causes entirely too much mental and nervous disturbance ever to become popular or justifiable as a medical measure." Nonetheless, he did "heartily endorse it as an additional punishment in certain offenses." Over the previous decade, Sharp had performed vasectomies on 456 men, all of whom became "of a more sunny disposition, brighter of intellect." Moreover, Sharp claimed, a sterilized patient "ceases excessive masturbation, and advises his fellows to submit to the operation for their own good." Later that same year, Sharp published an article in the *Southern California Practitioner* in which he described the "Indiana Idea" as a surgical procedure "by which we prevent people, of mental defect and transmissible physical disease from procreating without, in any way, endangering life or incapacitating them in their enjoyment of life, health, and pursuit of happiness other than loss of procreative power." After explaining how he had sterilized juvenile male and female calves to demonstrate that severing the oviduct or vas deferens had no effect on the animals' development, he concluded that the operation would "materially lessen the number of illegitimate children as well as decrease the population of our county poor asylums, almshouses and old ladies' homes."[68]

When medical doctors used the term *heredity* in the nineteenth century and in the first several years of the twentieth century, they understood it to mean something fundamentally different than biologists did. Heredity for medical doctors included both the biological inheritance that we receive from our parents as well as the environment in which we develop. There is no better analysis of nineteenth-century medical doctors' notions of heredity than Leila Zenderland's 1998 *Measuring Minds*. She explains, "Despite biologist August Weismann's controversial proofs in the 1880's that acquired traits could not be inherited, most physicians still believed the contrary."[69] Citing examples of the "Lamarckian legacies" that shaped medical doctors' assumptions about heredity, she offers the example of Martin W. Barr, the chief physician at the Pennsylvania Training School for Feeble-Minded Children and author of one of Pennsylvania's compulsory sterilization bills. He included as part of heredity the effects of pregnant mothers' "poverty, hard work, not infrequent intemperance, and many anxieties." In addition, "indulgence in petty vices, irresponsibility or consequent inability to attain success in life," he argued, was "almost sure to develop idiocy or imbecility in offspring."[70] While many biologists differentiated such influences on a person's development as environmental, rather than hereditary, nonetheless, medical doctors and some American biologists opposed interpreting Weismann's conclusions so narrowly as to exclude the ability of reforms like public education and temperance to improve the biological quality of a given population. As we will see in the next chapter, the question of the extent to which Weismann's work could inform public policy decisions regarding reform became an important part of Charles Davenport's solicitation for research funding after the turn of the century.

Medical Doctors as Advocates for Compulsory Sterilization Laws

Indiana enacted its compulsory sterilization law in 1907, and two years later Washington, California, and Connecticut followed suit. Within four more years, eight more states joined the list, while another four states passed compulsory sterilization laws that received gubernatorial vetoes. By 1913, nearly one-third of Americans lived in states that had compulsory sterilization laws.[71] In the midst of these new laws, American doctors offered one last burst of support for the use of sexual surgeries to solve social problems. In addition to Sharp's ongoing campaign in favor of compulsory sterilization, William T. Belfield, a Chicago medical doctor and specialist in diseases of the male urinary tract and sexual organs, campaigned throughout 1908 and 1909 in favor of laws that would enable physicians to perform involuntary sterilization on patients deemed mentally, physically, or morally defective. In a 1909 article, Belfield called on "intelligent people everywhere" to consider the eugenic value of vasectomies for defectives. Explaining that it was "not an iridescent dream," Belfield advocated the demise of the race of defectives that, he argued, was increasing in number at an alarming rate. Editors of the journal followed his article with comments on several other published endorsements of eugenic sterilization. Three years later, Belfield's written testimony would be included in one of the first constitutional challenges to a compulsory sterilization law. In *Washington v. Feilen*, the Washington State Supreme Court quoted Belfield's claim that a vasectomy was "less serious than the extraction of a tooth" in deciding that compulsory sterilization did not constitute cruel and unusual punishment.[72]

In January 1908, the editors of the *JAMA* came out in favor of compulsory sterilization as a solution to the many complicated problems it believed originated in biological inferiority. In "Race Suicide for Social Parasites," the editors posited that it was the civic obligation of physicians to make public pronouncements about such matters, because "in our professional capacity we can teach the public not how to punish but how to restrict crime by restricting the breeding of our criminals." Animal breeders, they claimed, demonstrate much greater wisdom in the care of their animals than did society in the care of their defectives, who were allowed to "breed more of their kind" and effectively rob society's own "worthy children" of the chance to adequately contribute to society. Ignoring claims from some that, freed from the threat of fathering children, vasectomy would encourage increased promiscuity among the unfit, the *JAMA* editors argued that defectives "seek pleasure rather than progeny," so a vasectomy would be sufficient to "prevent the transmission to offspring of their own hereditary taints, such as insanity and syphilis." The *JAMA* editors were encouraged by the recent passage of the nation's first sterilization law and Sharp's claim that over two-thirds of the 300 patients sterilized at the Jeffersonville Reformatory had requested the surgery. Demand among defectives for sterilization should, they concluded, remove "the only conceivable opposition to this method of protecting society—namely, the sentimental."[73]

The following year, the *JAMA* published a short news story that described how Illinois state legislators were debating a bill that would allow for the compulsory sterilization of the state's defectives and confirmed criminals. "There are doubtless many," the editors claimed, "who realize the necessity for some measure that will limit the output of ready-made potential criminals and defectives." Illinois Senate bill number 249, which was in committee in the spring of 1909, did not make it through the state legislature; in fact, Illinois was one of the few northern states that never adopted a sterilization law. There were, nonetheless, involuntary sterilizations performed in Illinois during the twentieth century. As was the case elsewhere, medical professionals simply did not need permission from state authorities to take away their patients' reproductive capacities.[74]

A few months after the second *JAMA* article appeared, Belfield published "Sterilization of Criminals and Other Defectives by Vasectomy" in the *Journal of the New Mexico Medical Society*, which the editors followed with reprints of several newspaper articles that supported the prevention of the procreation of certain classes. Belfield invited "intelligent people everywhere" to consider his claims that "natural criminals, imbeciles, insane, and epileptics" were especially fecund and that there were few legal constraints to "restrict the procreation by these irresponsible parasites on society." Some states, he explained, had passed laws requiring the sterilization of these classes of citizens, which obviously improved the "financial, moral and social health of every community" and demonstrated "true philanthropy." Belfield offered alarmist claims about the rapid increase in the number of defectives and praised the five states—Minnesota, Connecticut, Kansas, Michigan, and Ohio—that had passed laws to forbid the marriage of certain classes of unfit citizens. As beneficial as these laws were, they did not go far enough, and Belfield offered in their place castration, colonization, and vasectomy. Castration was simply too drastic "because it destroys the subject[']s sexual power; it unsexes a man." Colonization would work, but practically it failed in comparison to the potential benefits of the vasectomy operation. Whereas many other authors used the term "race suicide" to describe the rapid increase in the number of unfit citizens and the eventual demise of the human race, Belfield used it to describe what he hoped would happen to the race of defectives: "The average man . . . heartily approves this method of race suicide for criminals and other defectives, because of the obvious advantage to the community." His article was followed by stories from the *Chicago Evening Post*, the *Chicago Tribune*, and the *JAMA* that further described the threat posed by defectives and their offspring.[75]

At least one periodical, albeit a relatively obscure one, attacked Belfield's claims along with the doctors at the 1908 Chicago meeting who discussed the use of sterilization in the elimination of crime. William H. Houser, president of the Physicians Club of Lincoln, Illinois, wrote a scathing letter to the editor that was published in *Ellingwood's Therapeutist*. While appreciating Belfield's alarmist claims about the rapid increase in the number of defectives, Houser attacked the assumption that heredity caused their deficiencies. Instead, he asserted that "75 percent of all the crime in this country is caused either directly, or indirectly, by the drink habit." Saloons were, he asserted, "human butcher shops," and "if you really want to kill the crime microbe, all you will have to do is to put on the lid good and hard and seal it down hermetically" on the nation's saloons.[76]

Houser's attacks on Belfield and on compulsory sterilization laws had little apparent effect on the discussion. In fact, reports of the laws in regional and national medical journals helped bring the late-nineteenth-century advocates of sexual surgery into line with early-twentieth-century advocates of compulsory sterilization. Take, for example, J. Ewing Mears, a Philadelphia medical doctor who, in 1890, had presented a paper before the Philadelphia Academy of Surgery on treating hypertrophy of the prostate gland by cutting the spermatic cord. Nineteen years later, in 1909, he published "Asexualization as a Remedial Measure in the Relief of Certain Forms of Mental, Moral and Physical Degeneration" in the *Boston Medical and Surgical Journal*. In it, he argued that the procedure should be used on mental and physical defectives to prevent them from reproducing their kind. He used some of the harshest language found anywhere in the coerced sterilization literature about the "constant and perilous menace to the well-being and welfare of the human race" presented by the nation's "perverts and degenerates, idiots, imbeciles, epileptics and the vicious insane, as well as criminals of a certain type who, as a rule are the subjects of sexual perversions and abnormal indulgences." These defectives, he explained, "armed with the potentiality of propagating their kind, were as dangerous to the integrity of the community and state as the foe armed with weapons of warfare." Originally, Mears preferred ligating the spermatic cord, which resulted in the atrophy of the patients' testis, and he opposed castration because of the trauma it caused and the example of a London surgeon murdered by a patient he had castrated. However, after reading about Belfield's paper, Mears abandoned his advocacy of ligating the spermatic cord and adopted the use of vasectomy for male degenerates and tuberotomy or tuberectomy for female degenerates. Regardless of the operation used, Mears asserted that medical doctors, "members of our noble profession," had a responsibility to the public health and were promoters of the public good, and were therefore obligated to advocate ever-increasing use of coerced sterilization of the nation's defective citizens. His argument about the physicians' professional and civic obligations mirrored those presented by the editors of *JAMA* a year earlier.[77]

In 1911, four years after the first sterilization law was passed in Indiana and after a total of four states had adopted such laws, Henri Bogart appealed to fellow medical doctors to encourage their legislators to pass compulsory sterilization laws in his article "Sterilization of the Unfit—The Indiana Plan." Bogart's call to arms was published in the Kansas City–based *Medical Herald*, which was marketed to midwestern medical professionals. "To secure sterilization for your state, doctor, you will have to get out and hustle," Bogart wrote to his colleagues. Their expertise in medicine obligated them to treat patients and to train citizens about how compulsory sterilization laws could substantially improve society: "It is not enough for you to think that it is a good thing. Your legislator probably has never heard of it. You as a professional medical man, will be able to send him to the session intelligently ready to help your state into the forward movement." Offering concrete examples of why "sterilization of the unfit is humane to the last degree," Bogart quoted from letters describing epileptic women who produced children similarly afflicted as well as the story of "Blind Bill," an "illegitimate child of a degenerate family" who lived in a county poor asylum and "grew more and more bestial, and for many years was kept in a special house with a grated floor" that could be washed out like an animal's cage. Blind Bill's family had "furnished its quota of pauperism and criminals, such as one would expect, and the common report . . . was that his father was a brother to his mother." Bogart asserted that "there are such decadent families in every community" and concluded, "Doctor, you who read this paper, will you be up and doing? Will you lift your voice and pour out your influence to further this measure?"[78]

Again and again, in articles that advocated eugenic sterilization and in those that described the history of eugenic sterilization, authors emphasized the fact that "operations were relatively simple" and that they did not directly affect "sexual desire or performance."[79] Especially in comparison with castration, severing the vas deferens was a much less brutal procedure, both in terms of the operation itself and in terms of the effects. While castrated men became permanently impotent in the months following the procedure, men who received vasectomies were still fully capable of engaging in sexual intercourse; the operation merely sterilized them. In his 1902 article advocating vasectomy in the *New York Medical Journal*, Harry Sharp declared, "The strongest argument against the advocacy of castration has been that it practically destroys the future enjoyment of life, and that the knowledge of the patient that he is deprived of sexual power has a very depressing effect."[80] The vasectomy operation, on the other hand, allowed a man to fully engage in sexual activities and thus apparently did not hinder his abilities to enjoy life. The ability to contrast vasectomy with the much more brutal operation of castration allowed sterilization activists to appear to be far less cruel in advocating sterilization as a method for punishing or treating patients and inmates or for improving the quality of the next generation. Just as years earlier Orpheus Everts, William A. Hammond, A. C. Ames, and F. L. Sim had all contrasted castration with execution to demonstrate castration as a humane and respectful alternative to the more drastic alternative, early-twentieth-century sterilization advocates compared the vasectomy to castration to demonstrate their more enlightened position as compared to earlier advocates.[81]

Among the most authoritative advocates of sterilization outside the United States was Robert Rentoul, a British medical doctor and social reformer who supported the use of sterilization for certain mental defectives to reduce the costs of the rapidly increasing class. His 1903 *Proposed Sterilization of Certain Mental and Physical Degenerates: An Appeal to Asylum Managers and Others* offered justifications for the compulsory sterilization of defectives based on evolutionary notions. Arguing that "our asylums and like places are practically manufactories for degenerates" and offering Herbert Spencer's claim that "to be a good animal is the first requisite to success in life, and to be a nation of good animals is the first condition to national prosperity," Rentoul offered alarmist descriptions of the rapid increase in the number of asylum inmates. This increase, he asserted, was caused by England's efforts to overcome natural checks on population: "Think how Nature would reduce the number of lunatics to the smallest proportion, were she not so persistently and deliberately thwarted!"[82]

At the 1906 meeting of the British Medical Association in Toronto, Rentoul portrayed defective persons as "the most dangerous citizens . . . especially from the procreation standpoint." Using data from the English Lunacy Commissioners, Rentoul described the increase of insanity among Englishmen and the Scots. He advocated tubal ligation and vasectomy for simple degenerates, but "lunatics, epileptics, idiots, confirmed criminals and inebriates, and habitual vagrants" he believed should receive castrations, so that they would have "no sexual desire, no sexual power, and no power to impregnate." Rentoul justified coerced sterilization on the grounds that his proposal protected the "liberty of the degenerates" by allowing them "a right to live, to enjoy life, and, if possible, to become useful workers," because they would not have to be executed or incarcerated nor would they need to be prevented from marrying once they were properly sterilized.[83] That same year, he published *Race Culture: Or, Race Suicide? (A Plea for the Unborn)*, in which he described in great detail efforts by American lawmakers to limit the reproduction of degenerates through the passage of marriage laws and their efforts to pass compulsory sterilization laws. Coming a year before Indiana enacted America's first compulsory sterilization law, Rentoul's analysis concluded, "The Americans focus too much attention upon the mere ceremony of marriage; because it follows that if these degenerates do not marry they will still go on begetting degenerate offspring, and so cursing it with disease and a living death."[84]

New Allies Emerge

After the turn of the twentieth century, just as Sharp and other American medical advocates of compulsory sterilization were increasing their efforts to sterilize their patients and enact laws requiring it, members of a number of other professions took up the subject. Several professions devoted to criminal justice, mental health, and charity adopted sterilization as an acceptable method to prevent the production of citizens incapable or unwilling to support themselves as well as for the maintenance and therapy of state wards. Between 1896 and 1918, the American Association for the Study of the Feeble-Minded published the *Journal of Psycho-Asthenics*, which was marketed to the directors of institutions that housed the feeble-minded, epileptic, and insane. The journal featured claims by many prosterilization authorities such as Martin Barr, the chief physician at the Pennsylvania Training School for Feeble-Minded Children and author of the second state law that attempted to legalize compulsory sterilization. Barr wrote the Pennsylvania bill that would have made it "compulsory for each and every institution in the State, entrusted . . . with the care of idiots . . . to examine the mental and physical condition of the inmates." If the examination showed that there was "no probability of improvement of the mental condition of the inmate" and "procreation is inadvisable," the institution was authorized "to perform such operation for the prevention of procreation as shall be decided safest and most effective."[85] He was also president of the Association of American Institutions for Feeble-Minded in 1897, and he called, in his presidential address, for the passage of compulsory sterilization laws around the country.[86] Barr continued his campaign throughout his career and was still publishing on the subject in 1920. In "Some Notes on Asexualization," he offered a litany of justifications for castrating men and women in state hospitals as well as a considerable number of case histories drawn from his own files. Among them were descriptions of patients like "W.D.," who proved such "a pernicious influence among other boys" that at "eleven and one half years old he was castrated," which brought on a "marked improvement in every way." Also included was "E.B.," a woman Barr described as moral imbecile of high grade and a nymphomaniac, who was "vulgar, sexually exaggerated, untruthful, a thief and absolutely unreliable; yet has attractive manners and is rather good looking." She received an oöphorectomy when she was sixteen and "improved to such a degree that she is now out in the world and a great assistance to her mother."[87]

In 1905, as Pennsylvania's state legislators approved a compulsory sterilization bill that was eventually vetoed by the governor, the *Journal of Psycho-Asthenics* published articles and an editorial advocating the sterilization of defectives. For example, in "Is Asexualization Ever Justifiable in the Case of Imbecile Children," S. D. Risley, a Philadelphia physician, justified sterilization of state wards by calling them "an ulcer on our social tissue." Instead of relying on evolutionary and hereditarian rhetoric, Risley described the "downward momentum of the law of degeneration" and identified the sources of degeneracy, including alcoholism, opium or cocaine habits, epilepsy, mental alienation and disease, prostitution, pauperism, and crime. Asexualization, in the form of vasectomy, would serve a therapeutic purpose for society as a whole, Risley argued, in that it would prevent those who fell victim to degenerating influences from spreading their infection to the next generation.[88] In the same issue, the editors explained that American physicians are not "by any means unanimously in favor of the radical surgical treatment advocated by some, although the sentiment is rapidly becoming favorable to it when applied to cases of marked moral delinquency including habitual criminal adults." They also printed a translation of an article on European attitudes toward asexualization, which demonstrated that some sterilization operations were being done on defective patients throughout Europe and physicians' sentiments toward it were similar to their American colleagues. The editors concluded with a brief report by Martin Barr on the results of asexualization operations on eighty-eight boys, all sterilized prior to the passage of any state law authorizing it. Barr reported that "in every case there was marked mental and physical improvement, the children growing stout, and acquiring large frames."[89]

At least some welfare professionals became interested in the use of compulsory sterilization to improve the well-being of children in their charge. In 1910, Edwin A. Down, a physician and president of the Connecticut State Board of Charities, presented "The Sterilization of Degenerates" at the first annual meeting of the Connecticut State Conference of Charities and Corrections. Reporting on the activities of his committee and commenting on the state's recently passed compulsory sterilization law, Down asserted that any damage done to patients' health or civil liberties in the process of sterilizing them was of small consequence in comparison to allowing them to "contaminate the race without the least restriction." The law, he explained, was not intended to be punitive, and he described and rejected a series of complaints against compulsory sterilization. Down concluded by attacking "misdirected sympathy for unfortunates, which does not realize that the protection of the public is the first consideration, and which pities the criminal rather than his victims, the pauper rather than his posterity, and the tramp rather than the community he inflicts himself upon."[90]

Perhaps the most prolific advocates of sterilization among those professionals interested in welfare issues was Hastings H. Hart, a minister and director of the Child-Helping of the Russell Sage Foundation.[91] Throughout the early years of the second decade of the twentieth century, Hart published and spoke on the subject of the sterilization of defectives. In 1912, he presented two papers before the American Prison Association that detailed the growing number of states that had already passed compulsory sterilization laws and heralded the work and claims made by Harry Sharp. His second address, "The Extinction of the Defective Delinquent: A Working Program," was presented at the meeting and reprinted in *Survey* the following year. It offered a road map for improving the nation by eliminating the source of crime and degeneracy, and he included calls for the passage of legal commitment laws that covered patients in both public and private institutions, the creation of colonies for the custodial care of the feebleminded, a focus on the care and protection of feebleminded women, and an increase in the funds available for the care and treatment of degenerate citizens.[92]

Ultimately, the passage of compulsory sterilization laws, something that American physicians had supported for decades, did not occur until a coalition of other biomedical, welfare, and social science professionals joined the physicians' crusade. As states began adopting compulsory sterilization laws, American biologists joined the movement and offered something crucial to the argument in favor of the laws: they provided scientific authority to back the claim that feeblemindedness, immorality, and criminality were biologically based and inheritable. American medical professionals had for years claimed this was true, but they had no basis other than anecdotal evidence. The entrance of biologists into the public discussions about coerced sterilization provided formal authority for their claim that inferiority was hereditary. However, as we shall see in the next chapter, biologists had much more to gain by signing on to the campaign in favor of compulsory sterilization than the movement would ever gain from their participation.

Regardless of the movement's need for biologists' contributions, we cannot overlook the fact that agitation in favor of compulsory sterilization laws originated long before the emergence of Mendelian genetics or the coordinated efforts of American biologists to promote the American eugenics movement. The nation's physicians, seizing what they assumed to be their professional and civic responsibilities, had argued in favor of coerced sterilization of certain problematic individuals for decades before they were joined by other professionals. Their work and the influences that spawned their crusade predated the emergence of genetic explanations of social ills and medical ailments.

CHAPTER 2

Eugenics and the Professionalization of American Biology

American biologists arrived quite late to the discussions about coercively sterilizing those citizens who were presumed to carry hereditary defects, and, it turns out, they were among the last to leave. Nonetheless, their influence on the movement was significant because they provided scientific authenticity to the claims made by sterilization proponents, and they established that at least some human traits, including certain clearly undesirable ailments, were heritable. For biologists, participation in the discussion about compulsory sterilization was part of their interest in the broader American eugenics movement, and they were vital to the advancement of the eugenics movement in the United States. As influential as the biologists might have been when they entered the public discussion about compulsory sterilization, they gained much more from participating in the movement than the movement gained from them. Ultimately, their participation earned them significant social authority as well as funds to pursue basic scientific research on heredity and evolution.

Biologists provided the compulsory sterilization movement with formal justifications for the commonly made claim that some socially or medically undesirable traits were inherited. In turn, American biologists received tremendous financial support, respect from social and political authorities, and ultimately recognition as valuable social authorities. This brought them increased status and, more important for the development of the profession of biology in the United States, it helped secure substantial funding for their research. Biologists achieved this in the context of three organizations devoted to biological research: the Station for Experimental Evolution at Cold Spring Harbor (SEE), the American Breeders Association (ABA), and the Eugenics Record Office (ERO). All three owed their creation, administration, and ultimately much of their success to Charles Benedict Davenport, a biologist and perhaps the most influential man in twentieth-century American biology. From the turn of the century until his death near the end of World War II, Davenport profoundly influenced the development of the profession of American biology by demonstrating to American policy makers and to patrons of science the ways in which basic scientific research on evolution and heredity could ultimately improve the nation.

Charles Davenport and American Biology

Late-twentieth-century histories nearly universally award the title of founder of the American eugenics movement to Charles Davenport. Born in 1866 in Stamford, Connecticut, and raised in a family of New England educators and businessmen, Davenport earned a B.A. in civil engineering at the Polytechnic Institute of Brooklyn. After briefly working for the railroad survey, he began graduate work in zoology at Harvard, earning an A.B. and then, in 1892, a Ph.D. under E. L. Mark.[1] Davenport met his wife, Gertrude Crotty Davenport, while he was a graduate student. Like Davenport, she was a biologist, and she had a B.S. and M.A. in biology. He entered the job market at a time when economic conditions evaporated the few opportunities that existed for academic scientists in the United States. For most of the 1890s, Davenport scraped together a living while his wife scanned the obituaries in *Science* and wrote letters to dead professors' universities soliciting jobs for her husband.[2] Davenport made good on his wife's investment. By the turn of the century, the thirty-five-year-old biologist was the director of the summer school of the Biological Laboratory of the Brooklyn Institute of Arts and Sciences at Cold Spring Harbor, an assistant professor at the University of Chicago, and the author of thirty papers and five books on evolution, variation, development, and morphology. His entrepreneurial spirit, which had carried him and his family through the lean 1890s, was evident in his work at the University of Chicago.

At the 1901 meeting of the American Association for the Advancement of Science, Davenport offered his colleagues a prediction for zoology over the next hundred years, titled "Zoology of the Twentieth Century." He began by arguing that history could be employed to formulate predictions of the future and explained that scientific development always began with description and progressed to comparative activities. He described the nineteenth century as "the morphological century," as systematic zoology demanded careful anatomical studies that eventually gave way to comparative anatomy, and comparison became "a fundamental zoological method." Embryology, he argued, was likewise born a descriptive science that eventually gave birth to comparative histology and comparative physiology. The widespread acceptance of "the evolution doctrine" furthered this trend, and zoology had become "immensely more complex, due to its developing in many lines, and that the new lines are largely interpolated between the old and serve to connect them."[3]

Extending his history of nineteenth-century zoology into the twentieth century, Davenport foresaw three lines of advancement. First, biological scientists would continue to use old methods to study old problems. While he was careful not to "belittle the old subjects, even when pursued in the old way," Davenport declared that he "would wish to blot out" those zoologists "whose reckless naming of new 'species' and 'varieties' serves only to extend the work and the tables of the conscientious synonymy hunter." He predicted that systematists would continue to revise genera and families, anatomists would explain structures in greater detail, comparative anatomists and embryologists would better understand the relationships between animals, and cytologists would add to the knowledge of inheritance by their "study of centrosomes, asters and chromosomes." "All these subjects," he concluded, "have victories in store for them in the new century." The second development Davenport envisaged lay in the introduction of new methods for studying old subjects. As the nineteenth century faded into the twentieth, Davenport explained, "the descriptive method has developed into a higher type—the comparative; and of late years still a new method has been introduced for the study of processes—the experimental." Morphologists and cytologists, he claimed, would make great advances in the twentieth century by taking up experimental techniques. Davenport also predicted that future zoologists would abandon "the rough language of adjectives" and adopt quantitative and statistical methods for both research and description. The Linnaean system, he argued, was doomed and eventually would be replaced by a decimal system that delineated an organism's evolutionary relationship to other organisms as well as its habitat and special adaptations. Finally, Davenport predicted that new problems would be explored by new sciences, like comparative physiology and the study of animal behavior, which were both currently in their infancies and "hardly worthy of the name of a science." Each would enter "an era of precise, critical and objective observation and record" that would make them true sciences. Future zoologists would also achieve significant breakthroughs in their ability to control biological processes, such as growth rates, cell division, color, and sex. "The direction of ontogeny and of phylogeny will be to a greater or less extent under our control." Davenport also envisaged significant gains for the emerging science of animal ecology, which had long been "the pastime of country gentlemen of leisure." He chastised his colleagues for their disdain of animal ecology, saying, "When zoologists fully awaken to a realization of what a fallow field lies here this reproach will quickly be wiped out." While Davenport wrote only one paper on ecology during his lifetime, he was keenly interested in the subject, and two of his students, C. C. Adams and V. E. Shelford, became highly influential ecologists.

Davenport used the study of evolution to illustrate the course that he predicted science would follow in the twentieth century. He declared, "It seems to me that the signs of the times indicate that we are about to enter upon a thorough, many-sided, inductive study of this great problem [of evolution], and that there is a willingness to admit that evolution has advanced in many ways." Davenport believed, as did many other biologists of his day, that evolutionary scientists needed to depart from speculative methods and ask specific questions about variation, heredity, selection, and environmental influences. To this end, he predicted that evolution would be studied with "comparative observation, experimentation and a quantitative study of results," like the work done by the Englishman William Bateson. Davenport devoted over a third of his address to a detailed example, drawn from his own work, of how experimental and statistical methods would be brought to bear on evolutionary questions. Describing his work on *Pecten irradians*, a bivalve mollusk that inhabited the Cape Cod coast, he showed how he used statistical methods to study variation in living and fossilized *Pecten* shells.

Davenport concluded his forecast for twentieth-century zoology with a call for increased funding. As zoology became entwined with other specialties, future zoologists would be expected to have even broader bases of knowledge, and Davenport argued that zoology would need greater financial resources to attract quality students. He feared, "Our best students slip from our grasp to go into other professions or into commerce because we can offer them no outlook but teaching, administration, and a salary regulated by the law of supply and demand." For the United States to contribute its share to the advance of zoology in the twentieth century, Davenport believed that colleges and corporations would have to provide better financial support for the biological sciences. The substantial government funding for science that began during the 1940s was simply beyond his imagination, although his ongoing efforts to demonstrate that basic research would produce practical applications was vital to establishing the claim that the federal government ought to fund scientific research.

Davenport's interest in predicting the future for zoology was the product of his desire to promote experimental and statistical techniques to his colleagues, and he cleverly disguised his agenda as an enthusiastic prophecy for zoology in the twentieth century. We see this as much in his praise of experimental and quantitative analysis as we do in his calls for increased funding for zoological research. His professional aspirations, combined with the dominant position that Davenport assumed when he secured substantial funding from the Carnegie Institution of Washington (CIW) in 1903, led to the development of zoology along many of the same lines that he predicted in his address two years earlier. When he offered his forecast, Davenport was only three years away from the opening of SEE, which he directed for the first three decades of its existence. As director of SEE, Davenport hired young experimental biologists, supported their work, and arranged for the publication of their findings. His predictions for zoology in the twentieth century were largely correct because he worked tirelessly to make them come true.

Conspicuously missing from Davenport's predictions for zoology in the twentieth century was any advocacy for eugenics. Why did Davenport, perhaps the most renowned eugenicist in American history, fail to mention eugenics in his forecast? In 1901, he was still several years away from becoming a vocal advocate of eugenics. Davenport did not become interested in the subject until sometime around 1905, after his wife persuaded him that eugenics was a viable biological research program and after he became increasingly involved with the newly founded ABA, the first American organization to sponsor the investigation and promotion of eugenics.

Figure 2.1. Charles Davenport and Gertrude Crotty Davenport (1902). Charles Davenport Papers, American Philosophical Society.

Gertrude Crotty Davenport's role in her husband's career and her interest in eugenics have been almost completely overlooked by historians. Her papers, which have been absorbed into his, are found at the American Philosophical Association and offer some suggestions about her influence on Charles Davenport, especially on the development of his interest in eugenics. For example, in 1905 David Starr Jordan wrote to answer her questions about the "Tribe of Ishmael," a midwestern ethnic group generally associated with Romanian gypsies and a common research subject of American eugenicists.[4] Likewise, in a 1907 letter she wrote to B. K. Bruce, she explained, "Somewhat under the auspices of the Carnegie Institution I am making a scientific study of human inheritance. It has seemed to me that I would get the most reliable and obvious statistics from studying the behavior of strongly contrasting characteristics when brought in conflict as they are when different races of peoples intermarry." Therefore, she continued, she was collecting data regarding the qualities exhibited by children of interracial marriages and hoped that Bruce would help her secure some information. She concluded, "The investigation is a purely scientific one looking only for the laws of human inheritance if indeed it is possible to discover such laws."[5] Both of these letters precede Charles Davenport's interest in eugenics, contain research questions with which he later became involved, and suggest that his wife played as powerful of a role in his increasing interest in eugenics as she had in helping him secure his first academic position at the University of Chicago.

Davenport as Institution Builder

Davenport made the leap from University of Chicago professor to leader of the American biological science community shortly after the turn of the century when, after repeated applications to the trustees of the CIW, he secured money to develop the summer school at Cold Spring Harbor into SEE. He first approached the CIW in the spring of 1902, but received little response. A year later, he again appealed to them, and in his revised application, he argued that the recent rediscovery of Mendel's work on inheritance combined with DeVries's mutation theory made necessary "an experimental station for the study of evolution." After having visited several European research stations during the summer of 1902, he believed that "more than ever is the importance of an experimental station felt where quantitatively exact experiments in breeding through many generations can be conducted and from which material can be supplied to various specialists for cytological and biochemical investigation." He had already secured a grant from the Wawepex Society of Cold Spring Harbor for ten acres of land "situated on the sea, with abundance of spring water, on a fertile hillside adjacent on the one side to the largest freshwater fish hatchery and on the other to the largest marine laboratory of New York State." He planned to raise money from outside sources to erect a building on the land at a cost of $15,000 and asked the Carnegie Institution to agree "to maintain the Station for a period of 25 years at the rate of $7,500 per year."[6] His initial proposal was turned down by the board of directors in part because of an ongoing debate among members about "whether the CIW should fund research organizations or only individual researchers."[7] A month later, in April 1902, Davenport wrote letters to several prominent American biologists explaining his plan to expand the summer school into a permanent year-round biological research center. He wrote, for example, to Alexander Agassiz, son of Louis Agassiz and director of the Museum of Comparative Zoology, who had established and directed a marine laboratory at Newport, Rhode Island, from 1877 until his death in 1910, explaining that at his proposed station, "the attempts will be made to test the validity of specific characters to try to transform species by rearing them under changed conditions."[8]

There is nothing in Davenport's 1902 proposal to the CIW about eugenics, much less about compulsory sterilization. However, in a letter written a year later to Frank Billings, a CIW board member, physician, and colleague of his at the University of Chicago, Davenport made obvious his belief that the study of evolution could have significant practical application to human affairs. "Evolution," Davenport explained, "has replaced the idea that man is apart from the rest of creation having been made of a superior type by a special dispensation of the Creator by the idea of man's origin out of some thing lower by lawful, orderly processes that are still at work raising him to a more perfect manhood." Precisely how evolution worked, he observed, was a mystery: "We do not know the processes of evolution; they have never been studied." The need for exacting research was especially important given the opposing views of the two dominant explanations for evolution. On the one hand was Darwin's hypothesis, which claimed that changes in its environment would act directly upon the organism to alter it, and that this difference will be inherited. Advantageous changes would enhance the organism's survival, while disadvantageous changes would force its extinction. Davenport described how the "practically unanimous 'opinion' of naturalists, led by the sharp *logic* of Weismann is that this explanation is all wrong." The modifications wrought on organisms by the environment, they claimed, "cannot be transmitted to the next generation. The second generation begins at precisely the same level with the first." Deciding which of the two explanations is more accurate, Davenport argued, is vital to human affairs because Darwin's notion of evolution encourages the belief that social reforms can better humankind, while Weismann's position denies it. Davenport wrote, "You can improve a man by putting him in a good environment. His early training teaches him elementary manner and moral and the church continues and extends these teachings. Good books are placed in his hands, his ideas are raised, his imagination kindled, his ambition to do his best aroused. Schools show him how to make use of his powers and show him the direction in which he can work to the best advantage. Through all these influences a person born in the slums can be made a useful man." However, Davenport concluded, according to the "prevailing opinion," which favored Weismann's claims, "the influence of all this care bestowed on the individual is not inherited by his offspring. They begin at precisely the same level that he did and receive no dowry of a finer mental stuff from all his intellectual accumulations."[9]

In his 1903 letter to Billings, Davenport also demonstrated his belief that increased funding for basic scientific research in biology could aid policy makers in addressing one of the central problems of his day: what could and should be done with Americans of African descent? "We have in this country," he wrote, "the grave problem of the negro—a race whose mental development is, on the average, far below the average of the Caucasian." Could the "negro race" be elevated to the level of Caucasians, or was Weismann correct and would future generations of blacks have to "start from the same low plane and yield the same meager results?" Davenport concluded, "We do not know; we have no data. Prevailing 'opinion' says we must face the latter alternative. If this were so, it would be best to export the black race at once." The proposed station for the long-term study of evolution could address these questions, Davenport explained, offering conclusive evidence about both the processes of evolution and the best social policy based on science and in accord with nature. It is one of many examples of Davenport soliciting funds for basic scientific research with enthusiastic promises of its eventual usefulness in bettering society.[10]

On June 11, 1904, Davenport oversaw the official opening of SEE, which had come about with substantial funding from the CIW. The station was "richly budgeted and equipped" and the envy of the world's leading biologists as well as a "warm-weather watering hole for many able biologists."[11] It was, as Philip Pauly explained, "one of the dream projects of American academic biology."[12] As director of SEE, Davenport helped lead the development of biology in the United States during a time of explosive growth in both funding and knowledge, collecting money from public and private sources and publicizing the results of the station's work whenever possible. From its opening in 1904, through its development into the CIW's Department of Experimental Evolution in 1918, Davenport ran and grew the institution. SEE served as the foundation for American research into heredity, evolution, and eugenics, much of which ultimately became the science of genetics. In 1962, the institution was renamed the Cold Spring Harbor Laboratory and today is devoted to research programs in cancer, neuroscience, plan genetics, genomics, and bioinformatics.

Davenport as Researcher

Under the auspices of SEE and with financial support from, among others, the CIW, Davenport both oversaw and personally undertook breeding experiments with insects, fish, and cats, but had difficulty producing papers from the work because he was often unable to master the necessary breeding techniques.

Figure 2.2. Charles Davenport with one of the cats from his breeding collection. Charles Davenport Papers, American Philosophical Society.

Beginning in 1904, with the creation of SEE, he produced a number of articles on the inheritance of particular traits in a range of organisms, including color inheritance in mice, black wool in sheep, eye and hair color in humans, and a number of different characteristics in chickens and canaries, two animals he did have some success breeding. His work on mice, completed and published in the first years of the twentieth century, demonstrated that Mendelism, which had recently been rediscovered, could not entirely explain heredity. He was, nonetheless, among the first Americans to examine Mendelism, and in 1901 he published one of the first papers on the subject, "Mendel's Law of Dichotomy in Hybrids."[13]

The nineteenth- and early-twentieth-century physicians who first advanced the idea of coercively sterilizing the nation's defective citizens predicated their crusade on the assumption that socially and medically undesirable traits were in fact inherited. They could offer countless bits of anecdotal evidence about fecund families rich in destitute children, of masturbating idiots and sex-crazed child molesters, and about horrible diseases visited on generation after generation, but they needed from biologists an explanation of precisely how these unfortunate traits moved from one generation to the next. The rediscovery of Mendel's work in 1900 provided the compulsory sterilization movement with a potential answer to that question, and it offered to Davenport a way to leverage financial support from reform-minded patrons. Davenport quickly began research on Mendelism and within a few years recognized its potential value to the study of heredity and to the funding of scientific institutions. By the 1920s, due in no small part to Davenport's research and his institution-building activities, biologists had identified a number of human traits that were inherited according to the Mendelian ratio of 1:2:1, including eye color and blood type. They had also determined that a number of tragic human ailments were likewise inherited according to this ratio, such as hemophilia and Huntington's chorea. The rigid probability of Mendelism and the identification of several diseases as inherited in a Mendelian fashion empowered the advocates of compulsory sterilization laws, and it provided Davenport with a valuable tool for demonstrating the usefulness of basic scientific research to potential patrons.

Much has been made by historians critical of Davenport's advocacy of eugenics of the sloppiness of his scientific research, and it is true that he was not a particularly careful or meticulous researcher. The most thorough biography of Davenport, E. Carleton MacDowell's 1946 "Charles Benedict Davenport: A Study of Conflicting Influences," provides perhaps the best evaluation of his scientific research and publications. After praising Davenport's institution-building talents and his ever-youthful energy when it came to increasing the size and capacity of SEE, MacDowell explained that Davenport never permitted "himself to relax and enjoy life passively, or to mediate and give original ideas a chance to float up to his unconscious mind." For Davenport, "life was too full of action for pondering the meanings of his results, or for critical evaluation, or even exactitude." Methodological persistence was "unnatural and disciplined" for Davenport, and it appeared only in occasional and temporary bursts.[14]

Among the best examples of Davenport's worst attributes as a researcher was his 1908 *Inheritance in Canaries*, which MacDowell said gave "shocking evidence of speed too great either for consistent tables or for sound logic." The work is predicated on the question of whether evidence about heredity obtained through breeding experiments with domesticated animals could be "applied to feral species as they are evolving 'in nature.' " Davenport stated that the claim had little validity, something he had argued two years earlier in *Inheritance in Poultry*. Given Davenport's solicitation to the CIW for funds to create a breeding station that would allow researchers to address fundamental claims about heredity and evolution, allegations that domesticated animals could not provide accurate information about natural processes in the wild presented a serious threat to his ability to solicit funds for SEE. He concluded that the canary, which had been bred in captivity "for only about 250 years," demonstrated that "distinctive characters have arisen which behave in Mendelian fashion," including the color and the quality of the birds' crests. Thus researchers' findings in organisms that had long been domesticated were likewise evident in an organism that had only recently been domesticated.[15]

The shortcomings of Davenport's *Inheritance in Canaries* did not go unnoticed. A year after it was published, A. Rudolf Galloway, a Scottish expert on canary breeding, published "Canary Breeding: A Partial Analysis of Records from 1891–1909" in *Biometrika*. Motivated, he explained, by Davenport's work on the subject, Galloway analyzed his records to see "how far they agree with the conclusion in that paper." He found that they did not, and explained that Davenport's mistaken conclusions were due to his poor preparations for his experiments. "It would have been advisable," Galloway wrote, "for Davenport to have selected the original stock with much greater care." Davenport's subjects were haphazardly chosen from among a collection of breeding birds that had been bred "purely for song quite regardless of colour and crest, the two points concerning which the author wished to test Mendel's theories." He concluded that Davenport's work demonstrated that when studying "Mendelian phenomena as occurring in fancy varieties," it was absolutely necessary that a strict definition "of the characters under examination be made, and that their nomenclature, and behaviour under varying conditions, be thoroughly understood," something Davenport obviously had not done.[16] Davenport's response was printed several months later in the same journal, and it clearly demonstrated his frustration about Galloway's remarks.[17] The influence of Karl Pearson in Galloway's critique is likewise apparent, and it must be noted that Pearson harbored considerable resentment toward Davenport because of Davenport's adoption of Mendelism, which Pearson took as a personal attack.[18] Galloway's rejoinder was followed by an equally critical attack on Davenport's canary research by David Heron. Years later, Heron again attacked Davenport and his work; his later critiques left the pages of professional journals and found their way into the widely read *New York Times*.[19] It was just one in a number of incidents between Mendelians and biometricians that raged throughout the first decades of the twentieth century. It pitted Pearson and his colleagues, who claimed that evolution occurred through slow, continuous variations, against Davenport, DeVries, Bateson, and other advocates of Mendelism, who believed that new variations arose via mutations and were passed as discrete units from one generation to another.[20]

As Davenport began researching and publishing on eugenics after 1910, his limitations as a researcher again became apparent. Among the most harshly criticized of the publications that emerged from his eugenic research was his and Mary Theresa Scudder's 1919 *Naval Officers: Their Heredity and Development*, in which they explored the trait of thalassophilia, the love of the sea. The vast majority of the book consists of brief biographies of sixty-eight naval officers that focused on their personal traits and the promise that they had demonstrated as young men. Coming on the heels of World War I, Davenport explained that the United States had recently attempted to rapidly increase the size of its naval and military forces, and he posed the question, "What is the best method of selecting untried men for positions as officers?" Demonstrating his sincere interest in using basic scientific research to improve American public policy, Davenport explained how his analysis of the personality traits of successful naval officers uncovered the fact that three common traits were found in most of the men he studied: a love of the sea, a capacity for fighting, and a capacity for commanding or administering. He focused on the first of these three, and investigated the possible inheritance of thalassophilia, which he described as "apparently a specific trait to be differentiated from wanderlust or love of adventure." The sailors he interviewed at Sailors' Snug Harbor, a retirement home on Staten Island for retired U.S. seaman, described their "strong love for travel on the sea," and they showed no interest in travel on land. "That sea-lust is an inherited, racial trait," Davenport argued, "is demonstrated by its distribution among the races of the globe." The great nations, particularly those in the Middle East and Europe, regularly produced powerful navies, while the Africans, Chinese, Polynesians, and New Zealanders showed no interest in the sea. "Sea-lust, it must be conceded is a fundamental instinct, and a man who has it is as clearly differentiated from one who lacks it as a tern is differentiated from a thrush in its choice of habitat." He concluded that in selecting untried men for naval commissions, "advantage may well be taken of the assistance that is afforded by the facts of juvenile promise and family history." Unless thalassophilia appears in at least one side of a recruit's family, he should not be given a naval commission.[21] Davenport and Scudder's *Naval Officers* was not well received in its day, and it has since received considerable comment as demonstration of the foolishness of eugenics.[22] It is perhaps more appropriate to judge it in the context of Davenport's shortcomings as a researcher combined with his desire to demonstrate the applicability of basic scientific research in order to generate financial support for it.

The American Breeders Association

As influential as SEE was in the advancement of biology in the United States and in the study of genetics throughout the twentieth century, it did little to influence public interest in eugenics or advance laws that required the sterilization of defective citizens. However, in partnership with two other institutions, both of which were of special interest to Davenport, SEE became the scientific cornerstone to the American eugenics movement. The first of these institutions was the ABA, which was created only a few months after the grand opening of SEE. The second institution, the ERO, was a direct product of SEE and opened in 1910 under its administrative oversight.

At the annual meeting of the American Agricultural Colleges and Experimental Stations at the end of 1903, representatives gathered to create a new organization that would advance both the practical and the scientific aspects of heredity by inviting "the breeders and the students of heredity to associate themselves together for their mutual benefit and for the common good of the country and the world."[23] ABA officials asserted that the purpose of their association was to bring "practical breeders into closer touch with the scientists, and the scientists into a clearer knowledge of the practical problems of the plant and animal breeders."[24] The biologists who participated in the early years of the ABA sought an explicit connection between their research and its possible applications. Reports from the meeting described the relationship as beneficial to both practical breeders and biological researchers, and participants imagined that by finding useful applications for the young science of heredity, they were elevating the status of the biological sciences. A 1910 editorial in the *American Breeders Magazine* asserted, "Science is taking hold of the forces of heredity as it has hold of the forces of mechanics, and the Twentieth Century bids fair to be the century of breeding."[25]

Throughout the *American Breeders Association Proceedings* and the *American Breeders Magazine*, biologists and breeders alike heralded the new association for its potential to elevate basic scientific understandings of evolution as well as their applications in the form of plant and animal breeding. A 1913 editorial in the *American Breeders Magazine* titled "These Are Times of Scientific Ideals" claimed, "In bringing the practical breeders together with the scientists there has been created in this Association an atmosphere of enthusiasm and an environment which is fertile with inspiration and which is favorable to maintaining high ideals." The editors asserted that the ABA was both a practical and a scientific organization and had, in the first decade of its existence, encouraged advances in breeding as well as in the knowledge that underlie the processes of heredity. The editorial concluded with a quote from Thomas Volney Munson, a Texas horticulturalist and grape breeder: "I regard the American Breeders Association as the most important and influential agricultural association in America, and probably second only to the American Association for the Advancement of Science in promoting the general progress and welfare of the nation."[26]

Despite the excitement expressed by many participants about the ABA's potential to increase contact between breeders and biologists, there was at least some resistance to the new organization. For example, in the first volume of the *American Breeders Magazine*, Davenport explained that "to the scholastic biologist of our universities the work of the 'breeder' has for long been regarded with contempt. Although recognized as a department of commerce, it has been regarded in many quarters as the least dignified department, associated in mind with the cowboy, the stable boy, the 'hayseed,' the country jay, the peasant of Europe. 'What do you do at the meeting of the Association,' says my university colleague, 'inspect hawgs, pass around "pertaters" and show up your biggest ears of corn?' "[27] A 1913 editorial in the *American Breeders Magazine* suggested that breeders held similar feelings about scientists when it stated, "The time of aloofness, of the men in rural pursuits, from the larger constructive and coöperative activities of society is coming to an end."[28] In retrospect, the biologists benefited more from their contacts with the breeders than vice versa. Clearly Davenport, David Starr Jordan, and Vernon Kellogg, a Stanford biologist and among the most widely read advocates of Darwinian evolutionary theory, believed that they would benefit from direct communication with breeders, and most every biologist in the ABA thought that plant and animal breeders were a storehouse of information about the workings of evolution. It is not so obvious, however, what the breeders had to gain from contact with biologists; nonetheless, Davenport made many promises about the potential of collaboration. Decades later work such as that done by ABA member George Harrison Shull on hybrid corn would finally demonstrate the value of geneticists to plant breeders.

It is often claimed that the ABA's Committee on Eugenics, founded in 1906, was the first national eugenics organization. In his 1908 "Report of the Committee on Eugenics," however, the chairman of the committee, David Starr Jordan, correctly stated that it was not in fact the first of its kind: "The National Conference of Charities and Correction, the American Prison Association, and probably other similar organizations have considered for a number of years, and are now considering more actively than ever, various phases of the broad subject of eugenics."[29] The ABA's Eugenics Committee set four duties for itself: "(1) to investigate and report on heredity in the human race; (2) to devise methods of recording the values of the blood of individuals, families, peoples and races; (3) to emphasize the value of superior blood and the menace to society of inferior blood; and (4) to suggest methods of improving the heredity of the family, the people, or the race."[30] The committee was organized and supported within the framework of the ABA on the basis that "the principles of heredity seem to be common to all plants and animals, including man," and "the data derived from the study of one kind of organism will be of importance in the study of other organisms."[31] The 1908 "Report of the Committee of Eugenics" defined eugenics as "the consideration of the bioligical [*sic*] factors influencing the conditions and the evolution of man."[32] Jordan explained that because of the evolution of sympathy, humans were in danger of undermining "the factors which automatically purge the state of the degenerates in body and mind." This development of altruistic sympathy, which Jordan heralded as a positive step in human evolution, must be addressed through the organization of charitable activities so as to avoid a "national shipwreck." That is, human evolution left unchecked by rational planning could be ultimately destructive because increasing levels of sympathy would encourage activities that would erode the quality of the species.

In 1910, Davenport suggested that the ABA reorganize and elevate the Committee on Eugenics to a section, thus putting it on par with the ABA's Plant Section and Animal Section.[33] Jordan supported the change, and later that year the members voted to create the Eugenics Section. An editorial note in the *American Breeders Magazine* asserted that the change was proposed because eugenics was "growing so rapidly in its work that the sectional organization with committees, rather than the committee organization with sub-committees, would facilitate and strengthen its work."[34] Editorials in the *American Breeders Magazine*, later the *Journal of Heredity*, as well as articles in the *American Breeders Association Proceedings*, applauded the ABA's interest in eugenics and the potential that eugenics held for improving the quality of the human species. One article concluded, "the race will experience its greatest improvement and attain its greatest ultimate physical and intellectual development through eugenics."[35]

A number of ABA members expressed concerns about making eugenics a section of the ABA, and in a 1910 article Davenport described the reactions elicited from some of the association's members: "When told of the Eugenics committee of the [American] Breeders Association some of these inquirers can barely restrain an expression of disgust that human interests should thus be mixed up with those of domestic animals."[36] In order to avoid becoming "blind leaders of the blind," Davenport encouraged careful investigation into the laws of heredity and warned that "premature attempts at education will bring the whole business into deserved reproach." "Our greatest danger," he asserted, "is from some impetuous temperament who, planting a banner of Eugenics, rallies a volunteer army of Utopians, freelovers, and muddy thinkers to start a holy war for the new religion." He believed that the association of eugenics with these groups would make later, more "sensible" applications ever harder to achieve. An editorial comment in the same volume of the *American Breeders Magazine* also evidenced uneasiness about the enthusiastic promoters of eugenics. It differentiated the ABA members who supported eugenics from other eugenic proponents by stating, "The group of workers chosen by the membership of this Association to work at this problem will be sane, safe and conservative."[37] Another editorial asserted, "No subject brought up for general discussion and solution during recent time is fraught with more possibilities for good or bad than eugenics."[38] The daily press added to members' concerns when it reported on the association's interest in eugenics "with levity and ridicule."[39]

Vernon Kellogg also expressed concern that eugenicists secure an accurate understanding of the natural processes involved in evolution before they attempted to apply their knowledge. In "Man and the Laws of Heredity," published in the 1909 proceedings of the ABA's annual meeting, Kellogg asserted, "A considerable part of our science of eugenics is based upon the knowledge of the order of inheritance which we assume in our formulation of these laws of heredity." Then he asked, "Is this foundation a firm one?" Like Davenport, Kellogg worried that premature generalizations would lead to "unwarranted dogmatism, to insufficiently grounded law-making, more speedily rebuked or more severely checked by the knowledge-seekers themselves than in the field of science." Kellogg concluded, "Let a swiftly—generalizing Hæckel or a speculative Weismann lift his head too soon or too high, and lo! It becomes at once the target for the shafts of his whole world of scientific confrères."[40] With a similar tone, Frederick Adams Woods used a short article in the 1909 *ABA Proceedings* to warn, "Much danger and confusion may arise when any facts drawn from our knowledge of the lower organisms are, by analogy, made to apply to man."[41]

Charles Woodruff's "Prevention of Degeneration the Only Practical Eugenics," published in the *ABA Proceedings* in 1907, concisely outlined some of the problems inherent to transferring knowledge and techniques from plant and animal breeding to improve the human species by focusing on the role of artificial selection.[42] Woodruff explained how Luther Burbank's methods of artificial selection, based on choosing a few of the best individuals and destroying the rest, are obviously not useful in improving humans. Even if it were acceptable to "do such an unnatural thing as to select human mates instead of letting them do their own selecting," which characteristics would be most advantageous to increase? "In every way it is viewed," Woodruff asserted, "any suggestion towards compelling the young to marry except as their instincts direct, is unnatural, unscientific, and absurd."[43] Instead of attempting to direct the selection of partners, he concluded, eugenicists should discover and address the causes of degeneration.

The ABA's promotion of eugenics is an excellent illustration of the way in which American biologists integrated their activities as biological researchers and progressive reformers by popularizing applications that supported their political positions. The belief that properly informed people would make correct decisions had long been a central tenet of the progressive reform movement. Similarly, eugenicists in the ABA believed that, provided with the appropriate scientific information, most people would act in a manner most beneficial to the species. A 1912 editorial stated that the ABA's stance on eugenics had been "to learn the truth and allow the truth to be its own power."[44] Alexander Graham Bell, a member of the Eugenics Section and an ardent supporter of eugenics, claimed, "The mere dissemination of information concerning those conditions that result in superior or inferior offspring would of itself tend to promote the production of the superior and lessen the production of inferior elements."[45] In another article, he asserted that the "fine art of selective breeding under skilled hands will never be a factor in human development"; instead, "the enlightened will of the individual must be in the long run the chief factor in selection."[46]

Davenport made a similar call for public enlightenment in the 1910 "Report of Committee on Eugenics," in which he asserted, "As precise knowledge [about eugenics] is acquired it must be set forth in popular magazine articles, in public lectures, in addresses to workers in social field, in circular letters to physicians, teachers, the clergy and legislators." He concluded by asserting the importance of another tool widely used by progressive reformers and explained that once biologists aroused the public to the importance of eugenics, their conclusions "must be crystallized in appropriate legislation."[47]

A second example of biologists in the ABA supporting a progressive reform issue is found in their promotion of euthenics, the scientific study of the effects of the environment on human development. Members argued that heredity alone was not the basis for the improvement of humanity; like progressive reformers, these biologists asserted that through temperance, increased public sanitation services, and education of parents on matters of hygiene, scientists could "raise the racial standard."[48] Bell, for example, claimed that "the improvement of the human race depends largely upon two great factors, heredity and environment."[49] Likewise, in a 1909 article in the *American Breeders Association Proceedings*, Roswell Johnson argued that direct modifiability of the germplasm had correctly replaced the discarded inheritance of acquired characteristics in justifying "the improvement of the environment and the prevention of the individual's abuse of himself."[50] Humans were products of their heredity as well as the environments in which their inherited traits were nurtured or stunted. Thus, Roswell concluded, "the social reformer may well feel justified in claiming the support of the biologist." In other magazines, like *Popular Science Monthly*, ABA members published articles that asserted the importance of both heredity and environment in improving the quality of humans.[51] Such statements illustrated the ABA members' belief in the importance of both heredity and environmental influences on the quality of the nation's citizens and show strong connections between progressive biologists and the other progressive reformers, such as temperance crusaders and social hygienists. Reformers did not need Lamarckian notions of the inheritance of acquired characteristics to justify their work once biologists began to accept that the environment and heredity worked in concert with one another. However, while poor environments could allow or prevent beneficial traits from appearing, even the best possible environment could not produce characteristics not present in one's biological inheritance.

The decline of the ABA became increasingly apparent shortly after 1910, as serious financial difficulties forced the association to abandon plans for experimental farms and cooperative projects with state experiment stations. Additionally, biologists in the ABA grew increasingly dissatisfied with the association and found alternative institutions in which to pursue their professional activities. Barbara Kimmelman explains that "as the ABA's agricultural station scientists transformed themselves between 1903 and 1913 into recognized students of genetics, the diverse institutional commitments of the organization constrained and diffused the sharply felt hopes and professional goals of this powerful ABA constituency."[52] Among the most vocal of the biologists who criticized the ABA were George Harrison Shull and Raymond Pearl, who joined T. H. Morgan, Edwin Grant Conklin, Davenport, William E. Castle, Bradley M. Davis, E. M. East, R. A. Emerson, and Herbert Spencer Jennings to establish the journal *Genetics* in 1916.

By 1911, many of the biologists who were instrumental to the development of the Committee on Eugenics and later the Eugenics Section stopped participating in the ABA. Herbert Windsor Mumford and Elmer Ernest Southard, neither of whom were biologists, took over the ABA's Eugenics Section and radically changed its nature. At the 1912 meeting, they added a number of new committees, including the Committee on Heredity of the Feeble-Minded, the Committee on the Heredity of Insanity, the Committee on Heredity of Criminality, the Committee on Immigration, the Committee on Sterilization, and the Committee on Inheritance of Mental Traits.[53] As Kimmelman has described, after 1912 the ABA quickly lost its momentum as members invested increasing time in other organizations and as opportunities for those interested in eugenics emerged elsewhere. Many biologists had already withdrawn from the Eugenics Section either because they did not agree with its politics or, more likely, they no longer saw in it the benefits to their careers and to their profession that they once did.[54]

In 1913, the ABA changed its name to the American Genetics Association; likewise, the name of the association's publication was changed from the *American Breeders Magazine* to the *Journal of Heredity*. The editors explained that the name changes were necessary because "the word 'breeders' no long accurately described to the general public the purpose of the association, since in the public mind it was connected solely with live stock." Continued advocacy of eugenics—or more accurately, the basic scientific knowledge of heredity that the editors believed should provide a foundation for the eugenics movement—also motivated the name change. The editors explained, "The steady growth of eugenics, and the full recognition granted to this new and important science by the association, have further made a change of name desirable." Members still recognized an urgent need for active leadership on the part of scientists, the editors argued, lest the eugenics movement "be detached from its proper basis of genetics and be captured by sentimentalists and propagandists with slight knowledge of its biological foundation."[55]

The Eugenics Record Office

The second institution that Davenport helped create and that brought eugenics and the issue of compulsory sterilization to the public's attention was the ERO, founded in 1910. That same year, Davenport had taken over as chairman of the ABA's Eugenics Section, and he was "filled with plans for great developments, which would require much money." To secure the needed funds, he assembled a list of wealthy Long Islanders from *Who's Who*, which included the Harriman family. Edward Harriman was a railroad tycoon who had taken over the ailing Union Pacific in the 1890s and quickly made it profitable again. In 1899, he funded the Harriman Alaska expedition, which explored and cataloged the natural history of Alaska and helped establish several prominent American scientists' careers.[56] Harriman had died in September of the previous year, and his daughter, Mary, had been Davenport's student at SEE several years earlier, so Davenport approached the family about some of his ideas. He was not alone; within the first few months after Harriman's death, his widow received more than 6,000 appeals for charitable donations.[57] During the first two weeks of February 1910, Davenport dined with Mary Harriman, sent her two letters, and took her mother to lunch. On February 16, he wrote in his diary that he and Mrs. Harriman had agreed "on the desirability of [a] larger scheme," and she agreed to a substantial donation to the station. It was, Davenport famously wrote to himself, "a red letter day for humanity!"[58] The scheme that Davenport and Harriman worked out eventually totaled more than $500,000, a contribution equivalent to over $10 million today. In May 1910, Harriman purchased a seventy-five-acre estate with a large residence that was located next to SEE's campus. She donated it to the station, and Davenport opened the ERO there in October.

Davenport named himself director of the ERO, but he was not responsible for its day-to-day operations. For that, he hired Harry Laughlin, who would become the principal proponent of compulsory sterilization laws in the United States, perhaps in the world. Laughlin was a teacher in the agricultural department of the State Normal School in Kirksville, Missouri, who was interested in the subjects of animal breeding and heredity. After writing a letter to Davenport in 1907 about chicken breeding, he was invited to enroll in Davenport's summer course at the Brooklyn Institute for Arts and Science.[59] Laughlin considered the course "the most profitable six weeks" in his life.[60] The two men stayed in touch throughout 1908 and 1909, and Davenport had encouraged Laughlin to join the ABA.[61] After securing funds from the Harriman family and encouraged by Laughlin's enthusiasm about the potential of eugenics to synthesize reform and biology, Davenport offered Laughlin the position as director of the ERO.[62]

The ERO was vital to the growth of the American eugenics movement and, via Laughlin's vigorous campaigning, the increasing popularity of compulsory sterilization laws across the country. It provided, as Garland Allen explained, "both the appearance of sound scientific credentials and the reality of an institutional base from which eugenics work throughout the country, and even in Western Europe, could be coordinated."[63] The organization had two general purposes: to conduct research on human heredity and to educate the public and policy makers about the importance of both research and its practical application. The ERO wed basic scientific research with a justification for public funding for that research by demonstrating how it could direct public policy in the most effective ways. Laughlin's first report, published in 1913, detailed the ERO's activities: it

Figure 2.3. Harry H. Laughlin and Charles Davenport. Harry H. Laughlin Papers, E-1-3:12, Special Collections Department, Pickler Memorial Library, Truman State University.

served as a repository for eugenical records, studied human heredity, advised the public on the subject of fit marriages, employed and trained fieldworkers to collect eugenic data, and disseminated the eugenicists' research and conclusions.[64]

Laughlin's primary contributions to the American eugenics movement lay in his passionate advocacy of compulsory sterilization laws and his work as the "expert eugenical agent" for the U.S. House of Representatives, which helped bring about a series of immigration restriction legislation in the 1920s. His 1922 *Eugenical Sterilization in the United States* was his magnum opus. An encyclopedic collection of documents and commentary, the book chronicled the rise of compulsory sterilization laws, court challenges to them, and all the eugenical, surgical, and political aspects of sterilization in the United States. It is still the single best source for information about sterilization laws, court cases, and propaganda. The book concluded with a "Model Eugenical Sterilization Law," which served as both a guide for advocates of sterilization legislation and as a target for opponents to coerced sterilization, who began to emerge in the late 1920s. Laughlin's model law called for the sterilization of "socially inadequate" people. The classes of the socially inadequate included, he explained, the feebleminded, insane, criminalistic, epileptic, inebriate, diseased, blind, deaf, deformed, and dependent. A socially inadequate person is anyone "who by his or her own effort, regardless of etiology or prognosis, fails chronically in comparison to normal persons, to maintain himself or herself as a useful member of the organized social life of the state." It did not include people whose abilities were hindered by age, curable injuries, or temporary physical or mental illnesses. Laughlin garnered more respect and more citations than any other advocate of compulsory sterilization. He was routinely cited by legislators and journalists in discussions about sterilization laws, and his zealous promotion of eugenics earned him respect during his lifetime and scorn from historians, journalists, and activists throughout the latter half of the twentieth century.[65]

A Eugenicist, but Not an Advocate of Compulsory Sterilization

While the dozens of articles and short publications on heredity inform us about his interests in the subject, Davenport's most authoritative statement about his beliefs and his research on genetics and eugenics came with his 1911 *Heredity in Relation to Eugenics*. Devoted to Mrs. E. H. Harriman in recognition of the money she donated for the establishment of the ERO, the book was Davenport's attempt to apply "recent great advances in our knowledge of heredity" to humankind. Eugenics, Davenport explained in the first chapter, "is the science of the improvement of the human race by better breeding," and from a eugenic standpoint, the success of a marriage "is measured by the number of disease-resistant, cultivable offspring that come from it." The aim of eugenicists was to "improve the race by inducing young people to make a more reasonable selection of marriage mates; to fall in love intelligently." The study of heredity was therefore necessary to better understand how a propensity for disease was passed from one generation to another, and the public needed to be educated about that understanding.[66]

After describing how unit characters are passed from parent to offspring intact, rather than as a blend of the two parents' traits, Davenport detailed the "modern laws of heredity" in the form of the Mendelian ratio of heredity. "Before any advice can be given to young persons about the marriage that would secure to them the healthiest, strongest children," he wrote, "it will be necessary to know not only the peculiarities of their germplasms but also the way in which various characters are inherited." Studying family records collected by representatives of the ERO, Davenport and his colleagues determined that a number of human characteristics were inherited in Mendelian fashion from one generation to the next. Among these, according to Davenport, are eye, skin, and hair color; hair form; stature; body weight; and musical, artistic, literary, mechanical, and calculating abilities. Diseases, or at least propensities to diseases, were likewise heritable, including chorea, multiple sclerosis, cerebral palsy, and Ménière's disease. Davenport also addressed the potential heritability of vague conditions or abilities such as temperament, general bodily energy, and general mental ability. In several cases, such as the "inheritance of peculiarities of handwriting," Davenport stated that there was no clear or satisfactory evidence. In the case of general mental ability, which he recognized was "a vague concept" but nonetheless one in common use, he wrote: "It is hard to recognize a unit character in such a series any more than in human hair color. Nevertheless there are laws of inheritance of general mental ability that can be sharply expressed." He also stated unequivocally, "Two mentally defective parents will produce only mentally defective offspring." Criminality likewise had a strong hereditary component, so the "question whether a given person is a case for the penitentiary or the hospital is not primarily a legal question but one for a physician with the aid of a student of heredity and family history." While Davenport is often condemned by later authors who discuss the American eugenics movement for his claims about the heritability of ailments or characteristics that we today believe are not influenced by heredity, it is important to recognize the fact that Davenport was hesitant to claim the inheritance of many social traits. For example, in discussing pauperism, Davenport explained it was "mainly environmental in origin." There was, he asserted, probably also some organic explanations in the form of mental incompetence or "shiftlessness." Nonetheless, he still placed considerable emphasis on the role of the environment: "Education is a fine thing," he asserted, "and the hundreds of millions annually spent upon it in our country are an excellent investment." Religious teachers likewise "do a grand work and the value to the state of properly developed and controlled emotions is incalculable." Finally, "fresh air, good food, and rest," along with "cleaner milk, more air, and sunlight," were valuable in improving the population, especially in crowded cities.[67]

Given Davenport's conviction that a great deal of the responsibility for undesirable human traits, such as criminality, epilepsy, alcoholism, and disease, rested on heredity, it is natural to assume that he would have favored compulsory sterilization of those who had these conditions. Near the end of the book, under a section titled "The Elimination of Undesirable Traits," Davenport claimed that the practical question in eugenics was: "What can be done to reduce the frequency of the undesirable mental and bodily traits which are so large a burden to our population?" Surgical operations, he explained, could prevent reproduction by either destroying or locking up germ cells, and he had no doubt that the state had the power to operate on selected persons. Moreover, "there is no question that if every feeble-minded, epileptic, insane, or criminalistic person now in the United States were operated on this year there would be an enormous reduction of the population of our institutions 25 or 30 years hence." But, he asked, "is it certain that such asexualization or sterilization is, on the whole the best treatment?"[68] Davenport concluded that it was not.

It will perhaps surprise many people to learn that Davenport opposed compulsory sterilization laws; he did so for a number of reasons. First, he believed that the laws were based on inaccurate or incomplete scientific information about heredity, and he claimed he was "struck by the contrast between the haste shown in legislating on so serious a mater compared with the hesitation in appropriating even a small sum of money to study the subject." The science administrator's interest in securing additional funds for basic scientific research was obvious. In addition, he argued that the definitions of the classes of people targeted by such laws, especially the so-called feebleminded, were unscientific because they were unclearly defined. "Shall we sterilize or forbid marriage to all children," Davenport asked, "whose mental development is retarded as much as one year? That would include 38 percent of all children, and one of yours, O legislator!" Even in cases in which individuals were several years behind others of their same age, there was no certainty to the claims that a particular person so retarded in his or her development would necessarily produce similar offspring. Finally, Davenport worried about the social impact of sterilizing and releasing patients and inmates to mingle with the general public. The six states that had passed compulsory sterilization laws by 1911, when Davenport published *Heredity in Relation to Eugenics*, had all called for the use of vasectomy, which "does not interfere with desire nor its gratification but only with paternity." This, Davenport believed, was not necessarily an appropriate operation for certain citizens: "Is not many a man restrained from licentiousness by recognizing the responsibility of possible parentage? Is not the same of illicit parentage the fortress of female chastity?" Would not, he asked, some of the people sterilized "become a peculiar menace to the community through unrestrained dissemination of venereal disease?" Castration, in the case of rapists, may well be far preferable since it would serve as a punishment, a prophylactic, and a treatment for their inappropriate urges. Ultimately, Davenport opposed the passage of compulsory sterilization laws in favor of the segregation of the feebleminded throughout their reproductive years. Unlike sterilization or asexualization, segregation was reversible; that is, if under a positive environment the patient demonstrated progress, he or she could be released as a full member of society. Davenport recognized that segregation was, at least in the short term, a more expensive alternative to sterilization, but nonetheless believed that "there is reason to anticipate such a reduction in defectiveness in 15 or 20 years as to relieve the state of the burden of further increasing its institutions."[69]

In 1918, seven years after the publication of *Heredity in Relation to Eugenics*, Davenport reiterated his prosegregation, antisterilization position at the trial of Frank Osborn, a twenty-two-year-old man and a resident of a state institution in New York who had been ordered sterilized. On the stand, Davenport testified that New York's compulsory sterilization law had emerged out of studies performed by eugenicists. He also stated that despite the many critical statements made about notoriously degenerate families like the Jukes and the Nams, "there is to be found much of good in the most degenerate families known in our land." He concluded that, for various reasons, "he has not advocated the operation of vasectomy, and that in his opinion segregation of the sexes would be better."[70]

Davenport's proeugenics, antisterilization position was not uncommon; other prominent figures likewise supported eugenics while finding fault with coerced sterilization as a proposed remedy for the perpetuation of inherited disorders. Take, for example, Henry Goddard, the director of the Research Laboratory of the Training School at Vineland, New Jersey, for Feeble-minded Girls and Boys and perhaps the nation's best-known psychometrician. He concluded his 1922 *The Kallikak Family*, which demonstrated that a hereditary taint was passed through the generations of a family he studied, with the question, "Why isn't something being done about this?" After discarding the idea of a "lethal chamber" that would do away with those who possessed hereditary defects, he explored the notion of solving the problem by taking "away from these people the power of procreation" by simple sterilization. "The operation itself," Goddard explained, "is almost as simple in males as having a tooth pulled," and "the results are generally permanent and sure." Like Davenport, he rejected sterilization as a potentially useful way of dealing with hereditary defects and laid special emphasis on the negative social consequences that could arise from widespread sterilization of the unfit. "What will be the effect," he asked, "upon the community in the spread of debauchery and disease through having within it a group of people who are thus free to gratify their instincts without fear of consequences in the form of children?" While he concluded that the "feeble-minded seldom exercise restraint in any case," he also argued that "segregation and colonization is not by any means as hopeless a plan as it may seem to those who look only at the immediate increase in the tax rate." Just as did Davenport, Goddard would have preferred to segregate defectives in colonies rather than initiate a widespread compulsory sterilization program. In addition to allowing for unrestrained promiscuity, he argued that sterilizing the Kallikak family would have "deprived society of two normal individuals" who "became the first in a series of generations of normal people."[71]

While eugenicists like Davenport and Goddard expressed reservations about compulsory sterilization laws as a solution to tainted heredity, Laughlin clearly differed in opinion from them, and he ardently supported the adoption of compulsory sterilization laws. Precisely why Davenport supported Laughlin's campaign for the passage of compulsory sterilization laws is unclear; perhaps he valued the attention that it brought to the ERO and SEE. Maybe it simply was not a significant enough of a difference in opinion to compel Davenport to raise the issue. While they differed on the subject of coerced sterilization, they shared a common view on the issue of immigration restriction. Both were strong supporters of limiting immigration to only those individuals who could demonstrate that they brought with them a healthy hereditary constitution.

We cannot today assess Davenport's relationship with the American movement to coercively sterilize some citizens by simply evaluating his personal opinions on the subject. Even though he did not support such laws, his scientific work and his institution-building activities as well as his personal and professional support of Laughlin substantially encouraged the compulsory sterilization movement. In 1917, the CIW took over responsibility for the operating expenses and future growth of the ERO. With an endowment of $300,000 from Harriman, the ERO enjoyed a level of financial independence that other CIW departments did not have. Davenport retired as director of SEE in 1934, and the ERO closed five years later when the CIW withdrew funding for it. Even with its endowment, the ERO was dependant on the CIW for operating expenses. Laughlin died four years later after reportedly suffering from attacks of epilepsy. As Allen concluded, "There is an irony in the fact that epilepsy was one of the traits that Laughlin and other eugenicists had wanted to purify out of the population; now he and his career became the victims of that neurological disorder."[72]

Davenport's death was a final demonstration of his devotion to creating and developing institutions. The seventy-seven-year-old died in February 1944 of pneumonia, which he had contracted after spending several cold January nights boiling a whale's head in a giant cauldron. The whale had washed up dead on a Long Island beach, and Davenport began the process of rendering the carcass to extract its skeleton. His last contribution to a scientific institution was a massive orca skull that was to be hung in the newly established Cold Spring Harbor Whaling Museum. After his death, he was memorialized by E. C. MacDowell and Oscar Biddle, both of whom emphasized his tremendous contributions to the American biological sciences.[73] Today, a century after Davenport created SEE, the Cold Spring Harbor Laboratory continues as one of the most important biological research facilities in the United States. For better or worse, our memory of Davenport is intricately bound up with the history of the American eugenics movement and its many effects, including the coerced sterilization of tens of thousands of American mental health patients, prisoners, and welfare recipients.

An analysis of the career and influences of the two premier American eugenicists, Charles Davenport and his employee Harry Laughlin, demonstrates three important facts about their role in the American eugenics movement. First, issues of patronage and a sincere desire to develop the profession of biology played a crucial role in Davenport's advocacy of eugenics. The lack of any interest on his part in the subject until well after the turn of the century and the immediate link he made between research on eugenics and fund-raising from the Harriman family and the CIW demonstrate how he saw in eugenics a way for American biologists to secure funding for basic scientific research in evolution and heredity. Eugenics, along with the other aspects of agricultural breeding that the ABA pursued, was the practical application of the research for which Davenport sought financial support.

The second revelation that emerges from this analysis is the fact that Davenport's and other biologists' work on heredity played a crucial role in the campaign for compulsory sterilization legislation. While American physicians had long campaigned in favor of such laws, their arguments about the inheritance of socially undesirable traits were grounded in anecdotal evidence and common-sense notions that like begets like. The rediscovery of Mendel's work in 1900 and the wave of American biologists who took up research on the subject provided the movement with a scientific explanation of what physicians could only assume existed: a reliable estimate of the passage of traits from one generation to another.

Finally, this brief exploration of Davenport's and Laughlin's careers, as well as the history of the institutions they led and the organizations they joined, demonstrates that one need not be a rigid hereditarian to believe in the potential of eugenics and of compulsory sterilization laws to improve American society. Current critiques often dismiss the eugenicists for their adherence to hard hereditarianism, which allows us to naively believe that the scientific advances that led to the acceptance of the important role of the environment—of nature *and* nurture—in human development brought the demise of the American eugenics movement and the end of coerced sterilization. In fact, Davenport and his colleagues generally believed that heredity and environment worked hand in hand to produce individual qualities. They were supportive of progressive reforms like temperance and compulsory education because they believed that good heredity was meaningless without equally good environments. Correspondingly, the reemergence of coerced sterilization in this country or of a neo-eugenics movement does not require us to become rigid hereditarians; obviously something more than just our scientific explanations of heredity and development is necessary to prevent us from repeating these mistakes.

CHAPTER 3

The Legislative Solution

Over the last 125 years, physicians in at least thirty-seven states sterilized some of the citizens that they considered unfit, and most of these physicians had the imprimatur of their states' legislatures. After decades of efforts, advocates of coerced sterilization finally persuaded thirty-two state legislatures to enact laws that would allow physicians to sterilize mental health patients, the chronically ill, and certain criminals. The call for compulsory sterilization laws was part of the progressive movement that swept the nation shortly after the turn of the century, and it included efforts to limit the marriages of certain citizens, which, it was believed, would likewise control who had children. Judging from the rhetoric employed by legislators, they were most strongly motivated to pass these laws by anxieties about the sexual activities of some of their constituents and by a desire to save money by reducing the number of people the state would have to house in its prisons and the growing number of mental health facilities.

From Marriage Restriction to Compulsory Sterilization

Excluding Gideon Lincecum's 1855 and 1856 efforts, legislation to prevent the procreation of unfit citizens originated with state regulation of the marriage of people with venereal diseases and those judged to be feebleminded. Advocates of these laws worked under the assumption that by preventing undesirable people from marrying, they were likewise preventing them from producing children. Among the first of these was the 1895 Connecticut law that prohibited "marriage or intercourse where either man or woman is epileptic, imbecile, or feebleminded, and the woman is under the age of forty-five."[1] In his 1895 address as president of the American Bar Association, James C. Carter described the Connecticut law as "a novel one, designed apparently to prevent unhealthy progeny," and punishable by at least three years' imprisonment.[2] Carter supported the legislation, calling it a "practical deterrent" and celebrating its ability to protect "future generations from the evil operation of the laws of heredity" that heretofore required "the perpetual imprisonment of habitual criminals." The Connecticut law sought only to limit marriage and illicit sexual intercourse; it did not call for the sterilization of its targets. But by 1910, advocates of sterilization laws recognized that marriage and commitment laws were valuable precursors to compulsory sterilization legislation. They also recognized that "there is procreation among certain classes without marriage," so more strident interventions than marriage laws were ultimately necessary if they were to "provide for a better posterity."[3]

In the nineteenth century, state interest in the institution of marriage was motivated by gendered notions of morality that were tinged with obvious signs of racism. Legislators enacted laws that would prevent interracial marriages—miscegenation laws—intended to protect the honor of white women and prevent the mingling of the races. John Jackson Jr., a historian who has studied the relationship between American science, the law, and race, concluded that the nineteenth-century justifications for miscegenation laws developed eugenic rationales in the twentieth century. Pointing to the 1924 Virginia Racial Integrity Act, the strictest such law in the nation, he explained that "it was the first miscegenation law in the nation passed on a eugenics basis."[4] In similar fashion, the emergence of marriage laws intended to stop the spread of physical and social ailments evolved from nineteenth-century concerns about the spread of sexually transmitted disease from husband to wife to child, to twentieth-century eugenic concerns about hereditary basis of defects.

Between 1907 and 1937, two-thirds of the states passed compulsory sterilization laws, and the majority of them had prior laws that regulated the marriage of citizens declared feebleminded or diseased. Of the thirty-two states that passed a law compelling the sterilization of prisoners and inmates of state institutions, 87.5 percent of them had a preexisting law that prevented some citizens, depending on their mental or physical status, from marrying. In comparison, of the sixteen states that did not pass compulsory sterilization laws, only 43.75 percent had laws that prevented certain classes of citizens from marrying; in other words, states with marriage restriction laws were twice as a likely to pass compulsory sterilization laws.[5] Therefore compulsory sterilization laws emerged from the earlier marriage restriction laws, and the laws targeted the same groups of citizens.

The Tenacity of Compulsory Sterilization Law Advocates

In the United States, compulsory sterilization laws were not a short-lived fad. They were passed again and again in states across the nation, and they were passed several times within the same states over the course of many decades. When some courts overturned the laws in the 1920s because they violated any one of several constitutional protections, legislators returned with new laws that got around the courts' challenges. Take, for example, the state of Oregon, whose legislators passed its first compulsory sterilization act in 1911 only to have it vetoed by the governor. At their next meeting in 1913, they passed another compulsory sterilization bill, which was signed into law by the newly elected governor, but fell to the state's first referendum vote. Apparently, most Oregonians did not want the state to coercively sterilize its wards. Four years later, as citizens' attention was focused on America's entrance into the Great War, Oregon's legislators again passed a compulsory sterilization bill, which operated until 1921, when one of the state's circuit courts invalidated it. The next time Oregon's legislators met, they passed a compulsory sterilization bill that got around the problems identified by the court. The law operated unchanged for over a decade before legislators loosened requirements to allow increased numbers of wards to be sterilized. Recognizing the tenacity shown by advocates of compulsory sterilization as well as the preexisting marriage laws, one sees that many legislators had long been interested in curbing the fecundity of certain of the state's citizens, and they were willing to work for decades to make it happen.

Despite the large number of states that passed compulsory sterilization laws, the total number of American men and women coercively sterilized as a punishment for crime, as therapy for any of a number of social disorders, or to better the nation's gene pool was very low, just a tiny percentage of the nation's overall population. For example, in California, which had by far the largest number of sterilizations, less than 0.2 percent of the population was coercively sterilized in the twentieth century. Even in Delaware, which had the highest per capita sterilizations, the percentage of sterilized citizens was less than 0.3 percent. Nonetheless, at least 63,000 Americans were sterilized under the authority a series of laws passed in nearly two-thirds of the nation's states.

Histories of coerced sterilization in the United States explain that Indiana was the first state to legalize compulsory sterilization, which it did in 1907. This is quite true. However, Indiana was not the first state to seriously consider such a law, nor was its state legislature the first to pass a law allowing physicians to coercively sterilize prison inmates and asylum patients. In 1897, nearly a decade before Indiana's landmark sterilization law and the same year that Ochsner first published his idea of using sexual surgery to decrease crime in the next generation, Michigan's state legislators debated an act that would sterilize the state's feebleminded citizens by castrating men or performing ovariotomies on women. House Bill No. 672, introduced in 1897 by Michigan state congressman and physician W. R. Edgar, was "a bill to provide restrictions relative to persons, inmates of certain State institutions, that such inmates shall cease to be productive, providing rules and modes of procedure to restrict the propagation of their kind."[6] Titled "An Act for the Prevention of Idiocy," it would have required the appointment of a "skilled surgeon" to examine the "mental and physical condition of the inmates," and, "if procreation is inadvisable, and there is no probability of improvement of the mental condition of the inmate," after one year's incarceration, it legalized involuntary sterilization.[7]

Legal authorities reacted critically to the Michigan asexualization bill, describing it as an act "so radical, not to say revolting, that it should be thoroughly discussed."[8] The *American Lawyer*, the news journal of the American Bar Association, described the Michigan bill as the "first attempt to pass a law of this kind." Demonstrating the punitive, rather than merely eugenic, motivations for the act, the journal asserted that the state "may have the power to provide this mode of punishment for the commission of the crime of rape," but it doubted that it had the "right or constitutionality of a law to inflict such an operation upon a subject who has violated no law, and no assumption can be legally indulged in that any law will be violated." As we shall soon see, these claims were in sharp contrast to those made three decades later by Supreme Court Justice Oliver Wendell Holmes Jr., who used the example of compulsory vaccination to justify the coerced sterilization of Carrie Buck, her mother, and her daughter in the 1927 *Buck v. Bell* majority decision. The author of the *American Lawyer* piece concluded, however, "There is a wide distinction between the legal status of this question and that of compulsory vaccination."[9]

While the state's lawyers generally opposed Michigan's 1897 asexualization bill, its medical community appeared to support it. The *Michigan Law Journal* published a complete description of the bill along with eight position statements from area medical doctors and one legal expert. The bill's originator, W. R. Edgar, explained how his attention was called to the subject while he was a congressman and learned that 210 inmates were housed at the Michigan Home for Feeble-Minded and Epileptic and another 600 had applied for admission. "When they are once admitted they must ever remain a charge to the state, as under our present system few are ever cured or discharged." Edgar believed his asexualization bill would reduce the population pressure on the state's institutions as well as future costs to the state after reading descriptions of how castration aided "confirmed masturbators" at the Asylum for Idiotic and Imbecilic Youth in Winfield, Kansas, and by accounts of "the skoptzy's or 'whitedoves,' " a Russian religious sect that practiced male and sometimes female genital mutilations. Ultimately, he explained he was motivated by the eugenic promise "of bettering humanity," believing that the "light of the future" will demonstrate the appropriateness of coercively sterilizing the state's insane, criminalistic, epileptic, and feebleminded citizens.[10]

Edgar's claims were echoed by three other physicians, William M. Donald, David Inglis, and J. J. Mulheron, each of whom offered long and detailed accounts of the potential benefits to the state and the patients that would come with the castration of habitual criminals, epileptics, chronic masturbators, and the feebleminded.[11] Donald pointed to bills in Connecticut, Indiana, and "even remote Texas" that sought to protect future generations by regulating marriage and reproduction. Proponents of Michigan's bill cited the work of Cesare Lombroso and Max Simon Nordau. Nordau's recently published *Degeneration* described the increasing moral, social, and biological decadence of European society.[12] These proponents campaigned in support of the bill based on four predicted uses of asexualization in dealing with unfit citizens. First, they emphasized "the deterrent influence the fear of castration would have upon the mind of the average criminal," which "would be felt only by those criminals who are still able to control conduct if stirred by some powerful influence."[13] Second, they described the therapeutic effects of castration, drawing analogies to stallions and bulls, on whose "violent and quarrelsome disposition castration has a most happy effect, soothing irritability, calming passion, and quieting anger."[14] Inglis asserted, "The entire change of temper and disposition effected in the horse by castration is strikingly suggestive."[15] Castration, several authors claimed, would also relieve feebleminded chronic masturbators, who were "addicted to that pernicious habit, 'Onanism,'" of their debilitating habit, "thus stopping the drain to their system."[16] Likewise, Edgar suggested that people with epilepsy might experience beneficial results from castration, and the passage of the bill would allow doctors to perform the operation "without working with the parents to gain their permission, as is the case now." Without using the word *eugenic*, advocates of the bill described the benefits to society of castrating the insane who possessed a "distinct hereditary taint, for these cases may be depended upon to propagate during their periods of sanity and to leave their fatal taint upon their offspring."[17] Finally, the bill's proponents concluded that castration was suitable punitive action for "rapists and seducers of youth," and a remedial treatment to help "victims of their own unbridled lust and passion" from committing sex crimes.

In the earliest debates about compulsory sterilization bills, eugenic rationales played only a small part in the discussion. As was the case with the earlier advocacy among American physicians for coerced sterilization, improving the overall genetic quality of the nation's citizens was just one of several rationalizations put forth for the sterilization of certain defectives. Just as influential as eugenic justifications were claims about the therapeutic effects of asexualization on chronic masturbators and other sexual perverts as well as the fact that the operation would make them easier to manage within state institutions. The punitive nature of castrating convicted rapists, child molesters, and men convicted of committing homosexual activities was likewise a powerful argument in favor of the passage of such laws. Imagining coerced sterilization as merely a eugenic activity allows us to overlook the fact that many different groups of professionals supported the passage of these laws for many different reasons, and it overemphasizes the role of biologists in the sterilization of tens of thousands of Americans throughout the twentieth century.

The *Michigan Law Journal* included statements from several opponents to the asexualization bill, including four medical doctors and one legal expert.[18] Whereas proponents of the bill drew support from criminologists and philosophers, its opponents cited the role of the works of Charles Darwin and Alfred Russel Wallace in instigating the legislation, and they quoted several of Galton's claims.[19] They argued against the bill's inaccurate and problematic language, such as its loose use of the term *degenerates* and its emphasis on hereditary influences over environmental. One opponent suggested that "the laws of Nature are too potent for puny man to oppose," so a program of controlled reproduction would do nothing to improve the next generation.[20] Nonetheless, most of the opponents accepted that "for the criminal guilty of rape, bestial assault or incest, castration would seem a particularly fitting and deserved part of judicial punishment, and a public sentiment which, not without reason, would hold up its hands in horror at proposed castration of defectives (feeble minded, imbeciles, idiots and epileptics) would heartily approve, applaud, and support some such law as that so crudely hinted at here."[21] The only legal authority to offer an opinion in the 1897 *Michigan Law Journal* discussion was Clarence A. Lightner, a lecturer on medical jurisprudence in the Detroit College of Medicine. Like his legal colleagues outside the state, Lightner was critical of bill, concluding that the law would probably be held unconstitutional because police power "does not seem to include future generation[s]." It would take, he argued, "increased knowledge on the part of the public of the need of restraining the increase of the insane classes," and, "if no other means are found to be effectual, the asexualization of the insane may finally be upheld on this ground."[22] Opponents of the bill focused almost exclusively on the eugenics rationales for coercively sterilizing state wards, and a supporter of the bill could have easily granted all their objections and still been able to argue effectively that it was justified on therapeutic or punitive grounds.

Despite support from several of the state's medical doctors and legal professionals, Michigan legislators voted down the state's first compulsory sterilization law. The bill was adopted by the committee, which recommended its passage, and Edgar claimed that he had the necessary votes. However, "on the day it came up for final vote there was a very exciting time in the Senate, and many of the members, friends of the bill, were absent, and did not get in to vote upon it. As it was, there were 45 who voted for it. A few seeing it had not the necessary 51 votes to pass, changed their votes to No."[23] The official tally showed thirty-nine in favor of the bill and forty-five opposed.[24] Sentimentalists, Edgar claimed, arose in the days after the vote, opposing the bill on irrational grounds and preventing it from coming to a vote again. Aside from his obviously biased account, we have no way of knowing precisely how the vote concluded or why the bill was abandoned, because all the records relating to such legislation have been destroyed. In 1951, a fire at the State Office Building consumed 8,000 cubic feet of state records and 20,000 books, including all the records pertaining to Michigan's 1897 act to allow coerced sterilization.[25]

In 1905, eight years after the Michigan bill died, Pennsylvania's state legislators debated a bill "for the prevention of idiocy." It was prepared by Dr. Martin W. Barr, superintendent of the Pennsylvania State Training School, and would have made it "compulsory for each and every institution in the State, entrusted . . . with the care of idiots . . . to examine the mental and physical condition of the inmates." If the examination showed that there was "no probability of improvement of the mental condition of the inmate" and "procreation is inadvisable," the institution was authorized "to perform such operation for the prevention of procreation as shall be decided safest and most effective."[26] State legislators may have been encouraged to pass the bill by statements like those made by Isaac Kerlin, president of the Association of Medical Officers of American Institutions for Idiotic and Feeble-Minded Persons and superintendent of the State Training School for Delinquent Boys in Elwyn, Pennsylvania. In his presidential address of 1892, Kerlin had claimed that "the census of 1890 unmistakably points to a steady increase in the proportion of idiocy and imbecility to the general population," and pointed out that, in Pennsylvania, there had been a 22 percent increase in feeblemindedness among the native born and a 228 percent increase in feeblemindedness among those in Pennsylvania of foreign birth.[27] Thirteen years after that speech, the legal authority Kerlin sought was offered when both houses of the Pennsylvania legislature passed Barr's bill for the prevention of idiocy.

Pennsylvania's sterilization bill was never enacted because Pennsylvania governor Samuel Pennypacker refused to sign it into law. He returned the unsigned bill to the state senate with a message that described his rationale for vetoing it. Pennypacker argued that the bill was too loosely worded and that it legalized experimentation of humans. He said, in part:

> If idiocy could be prevented by an Act of Assembly, we may be quite sure that such an act would have long been passed and approved in this state . . . . What is the nature of the operation is not described, but it is such an operation as they shall decide to be "safest and most effective." It is plain that the safest and most effective method of preventing procreation would be to cut the heads off the inmates, and such authority is given by the bill to this staff of scientific experts . . . . A great objection is that the bill . . . would be the beginning of experimentation upon living beings, leading logically to results which can readily be forecasted.[28]

Despite his obvious concern to limit the ability of the state to abuse its wards, Pennypacker was not particularly popular, especially among the state's newspaper reporters. He treated newspaper correspondents in a cavalier fashion, which resulted in widespread ridicule of him in the papers. His ultimate response to journalists' criticisms was reminiscent of Gideon Lincecum's half a century earlier. At the end of his term he spoke at an annual reporters' dinner and was met with catcalls, whistles, and boos from the assembled newspapermen. He raised his arms for silence and loudly stated, "Gentlemen, Gentlemen! You forget you owe me a vote of thanks. Didn't I veto the bill for the castration of idiots?" His rejoinder "brought down the house and assured him a respectful hearing from there on."[29]

After failing to be enacted or signed into law in Michigan and Pennsylvania, a compulsory sterilization bill appeared in the Oregon legislature in 1907. Bethenia Owens-Adair, a local progressive reformer and longtime advocate of compulsory sterilization, had introduced the idea of sterilizing the state's inferior citizens through letters to the editor in the *Portland Oregonian* several years earlier. While many of the state's legislators approved of her idea, too few of them voted in favor of the bill. However, that same year Indiana's legislators did approve such an act and, unlike in Pennsylvania, the governor signed it into law.[30]

The Nation's First Sterilization Law

After failed attempts to enact a compulsory sterilization law in Texas, Oregon, Michigan, and Pennsylvania, proponents finally scored a victory in Indiana. Founded on the claim that "heredity plays a most important part in the transmission of crime, idiocy, and imbecility," it targeted state wards held in mental health hospitals. Legislators charged the administrators of "each and every institution in the state, entrusted with the care of the confirmed criminals, idiots, rapists, and imbeciles," to appoint two "skilled surgeons of recognized ability" who would "examine the mental and physical condition of each inmate." If, in the judgment of the appointed experts, a patient had "no probability of improvement" in mental and physical condition, "it shall be lawful for the surgeons to perform such operation for the prevention of procreation." The only limit on compulsory sterilization was cost; no patient's evaluation should cost the housing institution more than $3.00. The law operated from 1907 until 1921, when the state supreme court decided in *Williams v. Smith* that it violated procedural due process because it failed to give patients opportunities for hearings or the ability to cross-examine the medical professionals who had ordered their sterilizations.[31] The court did not take up the question of whether compulsory sterilization was cruel and unusual punishment, nor did it address the claim that the law effectively empowered the legislative branch to determine punishments independent of the judiciary.[32]

Six years later, Indiana's legislators passed Indiana Acts 1927, Chapter 241, which overcame the court's criticisms. The new law required a thirty-day notice prior to the sterilization operations, and it gave inmates and their guardians time to prepare an appeal if they so desired. It was never directly challenged in court, perhaps because most sterilization hearings were nonadversary proceedings.[33] Acts passed in 1931 and 1935 added to the law the requirement that physicians who admitted idiotic, imbecilic, feebleminded, or insane patients to public institutions had to certify to the court whether the applicants had poor genetic constitutions.[34] In 1937, a revision to the law allowed the governing boards of each state institution the final authority to order sterilizations of its wards.[35]

In 1949, just as the state was beginning work on a $20 million mental hospital, its sixth such institution, C. O. McCormick, a medical doctor from Indianapolis, published "Is the Indiana 1935 Sterilization of the Insane Act Functioning?" in the *Journal of the Indiana State Medical Association*. McCormick reported the results of a questionnaire he had sent to the superintendents of the five state hospitals of the insane in which he found that the hospitals had admitted nearly 1,700 patients in 1948 and that almost 10 percent of commitments were accompanied

TABLE 3.1
STERILIZATION LAWS IN U.S. STATES

States that sterilized citizens under a compulsory sterilization law (Date of enactment)	*States that sterilized citizens but never passed a compulsory sterilization law*	*States that never passed a compulsory sterilization law and have no record of coercively sterilizing citizens*
Indiana (1907)	Colorado	Arkansas
Washington (1909)	Illinois	Florida
California (1909)	Pennsylvania	Kentucky
Connecticut (1909)	Texas	Louisiana
Nevada (1911)	Ohio	Maryland
New Jersey (1911)		Massachusetts
Iowa (1911)		Missouri
New York (1912)		New Mexico
North Dakota (1913)		Rhode Island
Kansas (1913)		Tennessee
Wisconsin (1913)		Wyoming
Michigan (1913)		
Nebraska (1915)		
New Hampshire (1917)		
South Dakota (1917)		
Oregon (1917)		
Alabama (1919)		
North Carolina (1919)		
Delaware (1923)		
Montana (1923)		
Virginia (1924)		
Maine (1925)		
Utah (1925)		
Minnesota (1925)		
Idaho (1925)		
Mississippi (1928)		
Arizona (1929)		
West Virginia (1929)		
Oklahoma (1931)		
Vermont (1931)		
South Carolina (1935)		
Georgia (1937)		

by court orders for their sterilizations. However, there were only three reported sterilizations that year because the institutions lacked the funds to perform the operations or because, despite their best efforts, they did not have adequate legal authorization to perform the surgeries. McCormick concluded that the state's sterilization law needed to be "grossly amended" to include, among other things, the ability to coercively sterilize patients in private hospitals.[36] Instead of altering the law to expand the number of sterilizations as McCormick requested, Indiana's legislators repealed the law requiring routine entrance examination in 1955, and the last recorded compulsory sterilizations in Indiana occurred in 1963, when the state sterilized twelve of its wards.[37] That same year the *Indiana Law Journal* published a brief history of the state's compulsory sterilization laws and called for their reevaluation given new scientific claims about the relationship between heredity and mental illness.[38]

State legislators in Indiana were not the first to consider a compulsory sterilization bill, but they were the first to successfully implement such a law, and discussion of it by advocates of compulsory sterilization encouraged passage of similar laws in other states. Two years after Indiana's law was enacted, compulsory sterilization laws were passed literally from one end of the country to the other. Connecticut, Washington, and California all passed sterilization laws in 1909, and two years after that, Nevada, New Jersey, and Iowa joined them. By 1920, eighteen states had enacted compulsory sterilization laws, and throughout the 1920s ten more states passed similar laws. The last state to join the list was Georgia in 1937, by which time thirty-two of the then forty-eight American states had enacted a compulsory sterilization law (see Table 3.1).

Coerced Sterilization without a Compulsory Sterilization Law

While it is easy to imagine that the states that passed compulsory sterilization laws effectively allowed physicians the greatest possible authority over state wards, the fact of the matter is that the states that passed these laws effectively limited physicians' ability to sterilize as they saw fit by mandating court orders, the oversight of boards of authorities, and the establishment of particular criteria for the sterilization of state wards. Prison inmates, mental health patients, and welfare recipients in Colorado, Illinois, Pennsylvania, Ohio, and Texas were coercively sterilized without the permission or the oversight provided by the government in other states. Among these five states, sterilization advocates were most vocal in Illinois, where state senators considered their first compulsory sterilization bill in 1909. Promoted by the Chicago Society of Social Hygiene, it would have allowed for the sterilization of criminals and defectives. The editors of the *Journal of the American Medical Association* supported the bill, although they were critical of its allowance for castration: "There are doubtless many who realize the necessity for some measure that will limit the output of ready-made potential criminals and defectives, who, nevertheless, are strongly opposed to what they consider the barbarous practice of compulsory mutilation, and these will have little fault to find with vasectomy."[39] The following year state representatives considered a bill that would have authorized the sterilization of "feeble-minded, insane, epileptic, inebriate, criminalistic and other degenerate persons" in situations in which "the opinion of a committee of physicians" determined that they could "produce children with an inherited tendency to crime, insanity or feeble-mindedness."[40] According to a critical analysis of the bill published in *Illinois Law Review*, it was based on questionable theories by "Italian criminologists" who claimed that "criminal traits are transmissible by heredity."[41] In both cases, the bills failed to garner enough votes to pass.

In 1916, the *Journal of the American Institute of Criminal Law and Criminology* printed a draft sterilization law introduced in Illinois that targeted the "feeble-minded, insane, epileptic, inebriate, criminalistic, and other degenerate persons" who had "inferior hereditary potentialities" and were "maintained wholly or in part by pubic expense." The bill would have brought about the creation of a Eugenics Commission, appointed by the governor, to "examine the innate traits, the mental and physical conditions, the personal records, and the family traits and histories of all the prisoners, inmates, and patients of the county and state institutions." Pending the commission's findings, members would file a report to the circuit court of the county in which the patient lived requesting that the court authorize the sterilization by determining if the patient had the potential for "reproducing offspring who would probably, because of the inheritance of inferior or anti-social traits, become a social menace, or a ward of the state."[42] Like its predecessors, the state legislature never approved the act.

In the years between 1910 and 1920, amid debates over sterilization laws in the state legislature, Illinois also hosted a widely discussed controversy over the euthanasia, either active or passive, of defective babies. In 1915, Anna Bollinger gave birth to a baby boy who was diagnosed with multiple physical anomalies, including the absence of a neck and one ear and deformities of the shoulders, chest, anus, skull, leg bones, and digestive tract. The hospital's surgeon and chief of staff, Harry Haiselden, told the parents that surgery could correct the baby's digestive tract and potentially save his life, but urged them not to request the surgery. They agreed, and five days later the baby died. In subsequent press coverage of the incident, Haiselden admitted that he had permitted "many other infants he diagnosed as 'defectives' to die during the decade before 1915. And over the next three years, he withheld treatment from, or actively speeded the deaths of, at least five more abnormal babies." His admission, framed by him in the context of eugenics, initiated a controversy over eugenic euthanasia and was depicted in the film *The Black Stork*.[43]

For some, the Bollinger incident justified the passage of a eugenic sterilization law to regulate medical doctors' interventions. For example, in 1916 the *Journal of the American Institute of Criminal Law and Criminology* published a letter from W. F. Gray of Clinton, Illinois, in which he described "the horrible torture of this little one fighting for life." The lack of a sterilization law allowed "quacks and men of ill responsibilities to grow famous" for their vigilante efforts to improve the race. For Gray, the Bollinger incident indicated the need for the passage of a compulsory sterilization law to control physicians' activities: "sterilization will make itself felt for the great good of the race, and when people learn to understand it, it will meet with their favor."[44]

In addition to being home to medical doctors like Chicago's William T. Belfield, who aggressively campaigned in favor of compulsory sterilization laws throughout the first decades of the twentieth century, Illinois was home to Judge Harry Olson, chief justice of the newly established Municipal Court of Chicago and patron of considerable eugenic research.[45] Olson's court was divided into a number of separate divisions that focused on specific types of crimes or social problems, such as the Morals Court for Women, Boys' Court, and the Automobiles Speeders Court. He pioneered the integration of the legal and social science professions through the establishment of the Municipal Court's Psychopathic Laboratory in 1914, which provided a venue for the scientific study of delinquency and its cures.[46] Olson worked with Harry Laughlin, making Laughlin a eugenics associate for the Psychopathic Laboratory and publishing both his momentous 1922 *Eugenical Sterilization in the United States* and his essay on the fundamental principles underlying heredity three years later.[47] Olson believed that criminality was the product of a hereditary defect, which until recently had been kept in check by harsh environments. He explained in his president's address to the annual meeting of the Eugenics Research Association at Cold Spring Harbor that the "normals have cut their rate of reproduction and at the same time have actually invited defectives to multiply freely with a guaranty that their offspring will be coddled and nourished and protected and brought by every artificial means to an age when reproductive instincts will provide another generation." He called for segregating "defective delinquents in state controlled colonies where the protective environment they need can be created."[48]

Physicians in Pennsylvania coercively sterilized state wards without the legal sanction of the state. Isaac Kerlin, an early advocate of sterilization, had not waited for authorities to legalize eugenic sterilization. In 1889, he had obtained parental consent for the castration of a feebleminded inmate.[49] Three years later, in his presidential address to the Association of Medical Officers of American Institutions for Idiotic and Feeble-Minded Persons, he asked, "Whose State shall be the first to legalize oöphorectomy and orchiectomy for the relief and cure of radical depravity?"[50] In 1905, Pennsylvania's legislators had been the first to pass a compulsory sterilization act, but it was never enacted because the governor had vetoed it. Each year from 1911 through 1919, state legislators considered a compulsory sterilization bill, about half of which died in committee, and the other half failed to pass in one or the other house. After the 1905 veto, the closest the state ever came to passing a compulsory sterilization law was in 1921, when George Woodward, a medical doctor and state congressman from Philadelphia, introduced a bill that passed in the Senate by an overwhelming margin but, like its predecessor, was vetoed by Governor William C. Sproul. In his veto message, Sproul explained that he rejected the bill because if the state had the ability to sterilize its feebleminded, it likewise had the ability to sterilize all others who were "considered undesirable citizens in the opinion of a majority of the legislature." Moreover, he argued that the bill violated the Fourteenth Amendment's guarantee of equal protection because it targeted those feebleminded citizens in state institutions, but not those in private institutions or still at large.[51]

Even though it was the home state of the first medical doctor to advocate the sterilization of criminals, Gideon Lincecum, and many of the state's medical doctors aggressively advocated sterilization for its therapeutic, punitive, or eugenic effects from the 1880s onward, Texas never adopted a compulsory sterilization law. Nonetheless, physicians coercively sterilized at least some of the state wards in their care; for example, in 1864 a Belton, Texas, jury sentenced a black man convicted of rape to be castrated.[52] In addition to considering Lincecum's Memorial in the mid-nineteenth century, in 1907 state legislators debated a law that would have allowed for the castration of criminals in cases of rape and incest. The *Texas Medical Journal* reported that the bill would probably pass in the lower house and that the law would avoid constitutional challenges as cruel and unusual punishment if it was intended "not as punishment, but as a sanitary or hygienic measure to prevent a repetition of the offense, in the interest of public morals and race integrity, and the propagation of a race of sexual perverts—for the propensity on the part of a negro adult to rape a small white child is a perversion—and doubtless could be transmitted." The journal's editors concluded, "It will be a big step in the advance of civilization if we can get such a law in every State."[53]

In 1927, a compulsory sterilization bill was passed by the Ohio Senate, but failed in the House because its backer attempted to force the vote prematurely.[54] Twenty-five years later, just as most other states were ending their programs of coerced sterilization, the Muskingum County Court of Common Pleas heard the case of Nora Ann Simpson, a "physically attractive young woman, aged 18" with an IQ of thirty-six who had given birth to an illegitimate child. Simpson, according to the testimony of her mother and by her own admission, "has been sexually promiscuous with a number of young men since the birth of the child." Her mother petitioned the court to authorize her daughter's sterilization to prevent her giving birth to additional children for whom she could not care. Quoting Holmes's claim from the 1927 *Buck v. Bell* case to justify the state's authority to sterilize "those who already sap the strength of the state," Judge J. Gary ordered her sterilized because it was "necessary for the health and welfare of said Nora Ann Simpson."[55]

General Trends in the History of Compulsory Sterilization in the United States

In examining each state's participation in the movement to coercively sterilize some of its citizens, certain patterns in the adoption and application of sterilization laws become evident. Every state on the West Coast and most of the midwestern states passed compulsory sterilization laws, and both groups of states were among the earliest to adopt such laws and among the most aggressive in

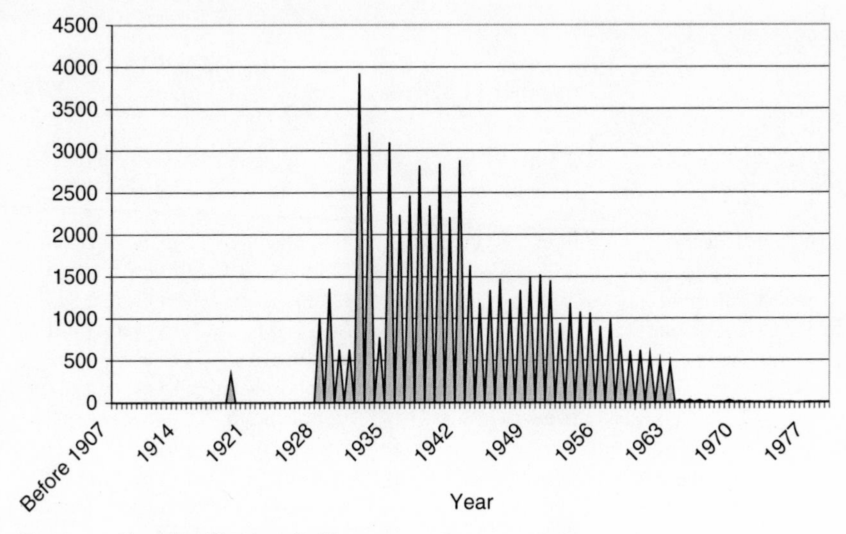

Figure 3.1. Total Sterilizations by Year, 1907–1980.

TABLE 3.2

TOTAL CONFIRMABLE COERCED STERILIZATIONS IN STATES THAT ENACTED COMPULSORY STERILIZATION LAWS, 1907–1983

Rank	*State*	*Number of sterilizations*	*Rank*	*State*	*Number of sterilizations*
1	California	20,108	17	Washington	685
2	Virginia	7,325	18	Mississippi	683
3	North Carolina	5,993	19	New Hampshire	679
4	Michigan	3,786	20	Oklahoma	626
5	Georgia	3,284	21	Connecticut	557
6	Kansas	3,032	22	Maine	326
7	Indiana	2,424	23	South Carolina	277
8	Minnesota	2,350	24	Montana	256
9	Oregon	2,269	25	Vermont	253
10	Iowa	1,910	26	Alabama	224
11	Wisconsin	1,796	27	West Virginia	98
12	North Dakota	1,029	28	New York	42
13	Delaware	945	29	Idaho	38
14	Nebraska	902	30	Arizona	30
15	South Dakota	789	31	Nevada	0
16	Utah	764	32	New Jersey	0

TABLE 3.3

PER CAPITA STERILIZATIONS IN STATES THAT PASSED COMPULSORY STERILIZATION LAWS

Rank	*State*	*Per capita sterilizations*	*Rank*	*State*	*Per capita sterilizations*
1	Delaware	1 in every 337	17	Wisconsin	1 in every 1,912
2	Virginia	1 in every 453	18	Montana	1 in every 2,309
3	California	1 in every 526	19	Maine	1 in every 2,803
4	North Dakota	1 in every 602	20	Mississippi	1 in every 3,190
5	Kansas	1 in every 628	21	Washington	1 in every 3,473
6	Oregon	1 in every 670	22	Oklahoma	1 in every 3,568
7	North Carolina	1 in every 678	23	Connecticut	1 in every 3,604
8	New Hampshire	1 in every 785	24	South Carolina	1 in every 7,643
9	South Dakota	1 in every 827	25	Alabama	1 in every 13,668
10	Utah	1 in every 902	26	Nebraska	1 in every 14,995
11	Georgia	1 in every 1,049	27	Idaho	1 in every 15,490
12	Minnesota	1 in every 1,269	28	West Virginia	1 in every 20,465
13	Iowa	1 in every 1,372	29	Arizona	1 in every 24,986
14	Vermont	1 in every 1,493	30	New York	1 in every 353,100
15	Indiana	1 in every 1,623	31	Nevada	No sterilizations
16	Michigan	1 in every 1,683	32	New Jersey	No sterilizations

Note: Per capita calculations are based on the total number of sterilizations as described in Table 3.4 and the U.S. Census data for 1930.

applying them. Compulsory sterilization laws were very popular among the states in which the progressive movement was especially strong, which explains why the states in the Midwest and West practiced coerced sterilization so often and for so long (see Table 3.2).[56] These were also the states that had the highest per capita number of total sterilization, with California having by far the most. Most northeastern states passed compulsory sterilization laws as well, but the per capita numbers of sterilizations in most of these states were relatively low, with the exception of Delaware, whose tiny population skews its per capita number of sterilizations (see Table 3.3).

In contrast to the strictly hereditarian language found in the original Indiana law and in the early laws of many other states, some states that adopted compulsory sterilization laws offered both hereditary and environmental justifications. In discussions about requiring the sterilization of state wards in Kentucky during the 1930s, at least one proponent of a compulsory sterilization law mingled hereditarian justifications for sterilization with environmental reasons, arguing that by sterilizing the mentally deficient, Kentucky authorities would prevent children from being "born in an environment entirely unsuited for the development of a normal personality."[57] The eugenic value of such a decree is unclear, but it obvious that legislators believed that defective parents could pass along

Table 3.4

Total Coerced Sterilizations by State, 1921–1980

State	To 1921[1]	To 1930[2]	To 1933[3]	To 1942[4]	To 1945[5]	To 1950[6]	To 1955[7]	To 1960[8]	To 1965[9]	To 1970[10]	To 1975[11]	To 1980[12]
Alabama	0	44	131	224	224	224	224	224	224	224	224	224
Arizona	0	0	20	20	20	21	21	30	30	30	30	30
California	2,558	6,787	8,504	15,220	17,399	19,423	19,937	20,023	20,108	20,108	20,108	20,108
Connecticut	27	200	338	426	490	515	532	547	557	557	557	557
Delaware	0	278	296	623	685	802	859	880	945	945	945	945
Georgia	0	0	0	136	360	803	1,961	3,012	3,284	3,284	3,284	3,284
Idaho	0	0	13	14	14	14	31	33	38	38	38	38
Indiana	120	123	217	1,129	1,360	1,889	2,197	2,378	2,424	2,424	2,424	2,424
Iowa	49	57	94	439	597	1,056	1,574	1,773	1,910	1,910	1,910	1,910
Kansas	54	657	976	2,583	2,851	3,001	3,025	3,025	3,032	3,032	3,032	3,032
Maine	0	12	41	212	218	238	278	318	326	326	326	326
Michigan	1	388	1,083	2,263	2,573	3,070	3,462	3,663	3,786	3,786	3,786	3,786
Minnesota	0	388	693	2,010	2,191	2,219	2,285	2,339	2,350	2,350	2,350	2,350
Mississippi	0	0	12	536	554	596	596	652	683	683	683	683
Montana	0	60	81	203	213	249	256	256	256	256	256	256
Nebraska	155	308	unknown	408	632	704	829	871	902	902	902	902
New Hampshire	0	61	165	436	511	571	658	679	679	679	679	679
New York	42	0	42	42	42	42	42	42	42	42	42	42
Nevada	0	0	0	0	0	0	0	0	0	0	0	0
North Carolina	0	3	46	1,199	1,605	2,401	3,663	5,051	5,993	5,993	5,993	5,993
North Dakota	23	39	93	575	684	807	911	986	1,029	1,029	1,029	1,029
Oklahoma	0	0	0	525	553	553	625	626	626	626	626	626
Oregon	127	650	882	1542	1,678	1,853	2,039	2,182	2,269	2,269	2,269	2,269
South Carolina	0	0	0	42	77	88	128	252	277	277	277	277
South Dakota	0	61	139	617	653	748	775	787	789	789	789	789
Utah	0	79	85	264	395	561	700	748	764	764	764	764
Vermont	0	0	30	221	232	251	252	253	253	253	253	253

(*Continued*)

TABLE 3.4

(CONTINUED)

State	*To 1921*[1]	*To 1930*[2]	*To 1933*[3]	*To 1942*[4]	*To 1945*[5]	*To 1950*[6]	*To 1955*[7]	*To 1960*[8]	*To 1965*[9]	*To 1970*[10]	*To 1975*[11]	*To 1980*[12]
Virginia	0	368	1,333	4,227	4,780	5,581	6,485	6,994	7,185	7,283	7,313	7,325
Washington	1	9	20	685	685	685	685	685	685	685	685	685
West Virginia	0	0	1	47	48	48	62	98	98	98	98	98
Wisconsin	76	305	492	1,219	1,467	1,658	1,764	1,772	1,796	1,796	1,796	1,796
Cumulative totals	3,233	10,877	16,066	38,087	43,791	50,666	57,177	61,540	63,701	63,799	63,829	63,841

Sources:

[1] Legal and Socio-Economic Division of the American Medical Association, "Reappraisal of Eugenic Sterilization Laws," which cited Laughlin, *Eugenical Sterilization*, 96; Robitscher, *Eugenic Sterilization*, 118–119; Whitney, *Case for Sterilization*, 302; Reilly, *Surgical Solution*, 97.

[2] Gosney, *Collected Papers on Eugenic Sterilization in California*.

[3] Reilly, *Surgical Solution*, 97; Whitney, *Case for Sterilization*, 302.

[4] Reilly, *Surgical Solution*, 97; Schmiedeler, *Sterilization in the United States*.

[5] Robitscher, *Eugenic Sterilization*, 118–119.

[6] Robitscher, *Eugenic Sterilization*, 118–119; Legal and Socio-Economic Division of the American Medical Association, "Reappraisal of Eugenic Sterilization Laws."

[7] Robitscher, *Eugenic Sterilization*, 118–119.

[8] Robitscher, *Eugenic Sterilization*, 118–119; Legal and Socio-Economic Division of the American Medical Association, "Reappraisal of Eugenic Sterilization Laws."

[9] Kindregan, "Sixty Years of Compulsory Eugenic Sterilization," 142.

[10] Robitscher, *Eugenic Sterilization*, 118–119.

[11] Robitscher, *Eugenic Sterilization*, 118–119.

[12] Robitscher, *Eugenic Sterilization*, 118–119.

problems to their children directly, through genetic inheritance, or indirectly, through social inheritance. Iowa adapted its sterilization law in 1946 to include a provision that would allow state authorities to sterilize anyone whose child they believed would probably become a ward of the state. Authorities interpreted this provision as justifying the sterilization of parents who lacked the ability to provide the proper environment for their child and would thus produce poorly raised children.[58] Montana's 1969 revision to its sterilization law empowered mental health authorities to sterilize anyone who they deemed would be "unable to adequately care for or rear such offspring without the likelihood of adverse effects on such offspring caused by such environment."[59] Six years later, Utah's legislators passed a law requiring that the courts, in reviewing an order to sterilize a mentally handicapped state ward, consider, among other things, the likelihood that the ward "could adequately care and provide for a child."[60]

Especially given the widespread assumption that eugenics and racism were closely intertwined, perhaps the most surprising fact that emerges from a state-by-state analysis of compulsory sterilization laws is that there were relatively few sterilizations in the Deep South. Of the six founding members of the Confederate States of America—Alabama, Mississippi, South Carolina, North Carolina, Florida, and Georgia—only North Carolina and Georgia had high per capita numbers of sterilization. South Carolina and Georgia were the last two states to pass compulsory sterilization laws, and Florida never passed a eugenically based compulsory sterilization law, although at the end of the twentieth century it did join the movement to coercively sterilize some convicted sex criminals. There are two reasons for the low incidence of compulsory sterilization in the South: first, sterilization laws in most states were adopted as part of the broader progressive movement. This explains why states like North Dakota, Kansas, and California had high per capita numbers of coerced sterilizations as well as why states in the South generally did not. The second reason compulsory sterilizations were not widely performed in the American South lay in the low number of institutions in most of these states. Few southern states invested the significant amounts of money necessary to establish mental health facilities during the first half of the twentieth century. Lacking these institutions, compulsory sterilizations were not performed even when the laws were adopted.[61]

California was by far the most aggressive among all the states that sterilized its degenerate citizens. The state's 1909 sterilization law authorized the superintendent of the State Home for the Feebleminded and the superintendents of state hospitals to sterilize their patients and inmates, which they began doing the following year.[62] In 1913, state legislators replaced the original law with one that required an explicit eugenic rationale for sterilization, making necessary the determination that the patient be afflicted with "hereditary insanity or incurable chronic mania or dementia."[63] Four years later, legislators amended the law to make the target those with "mental disease which may have been inherited and is likely to be transmitted to descendants."[64]

During the first ten years of the law's operation in California, fewer than 200 patients a year were sterilized. That number increased dramatically in 1917, averaging well over 300 a year until 1926, when it broke 500 for the first time. It stayed at between 500 and 800 per year until 1942, when it spiked to 1,333, before dropping back below 500 a year throughout the 1940s. During the 1950s, it was rare for more than two dozen patients to be coercively sterilized in one year. In total, more than 20,000 Californians were coercively sterilized under the state's sterilization laws, and in March 2003 Governor Gray Davis officially apologized for his state's history of sterilization.[65]

Due in part to the variety of materials available on compulsory sterilization in the California, historians of eugenics have been especially productive in chronicling and analyzing the social, medical, political, and scientific influences that helped make the state's professionals so enthusiastic about eugenic sterilization. Among the most recent of these is Alexandra Minna Stern's *Eugenic Nation*, in which she argues that previous analyses of the subject have suffered by strained connections to the Nazis and the domination of an East Coast–centric viewpoint.[66] Stern's argument, combined with the exceptionally high number of sterilizations in California and the very different contexts for the state's eugenic aspirations, makes it clear that any singular history of compulsory sterilization laws will always suffer from overgeneralization. Given the decentralized nature of public health in the United States, the regional influence of progressive reformism, and the various motivations for coercively sterilizing state wards, generalizing about the root causes and the effects of compulsory sterilization laws is difficult. It can be said that from the 1890s through the 1930s, various combinations of eugenic, punitive, and therapeutic justifications were offered for the passage of these laws. In practice, especially between 1920 and 1930, the poor were disproportionately impacted because they were much more likely to become wards of the state than were wealthy citizens. The advocates of these laws were well-meaning, even if the language that they employed and the authority that they gave to the state were at times quite brutal, and they honestly believed that they were improving citizens' lives, at least in the aggregate. Finally, the enactment of compulsory sterilization laws across the country and over the course of three decades demonstrates the interstate influence of American professional groups, who effectively convinced legislators around the nation to coercively sterilize some of their state's citizens.

Court Challenges to Compulsory Sterilization Laws

Once a state enacted a compulsory sterilization law and authorities chose to sterilize one of their wards, the only way for the patient or prisoner to avoid the operation was to petition the courts. In doing so, many of the laws themselves came under judicial review. Authorities in some states avoided court challenges by securing consent from legal guardians, which effectively negated the abilities of the patients and inmates from challenging the order for their sterilizations. Take, for example, North Dakota's 1913 compulsory sterilization act, which included in its purview all habitual criminals, along with the insane, idiots, defectives, and rapists.[67] Within a year of its passage, Dr. W. M. Hotchkiss, superintendent of the State Hospital for the Insane in North Dakota, stated in a letter to the American Institute of Criminal Law and Criminology's committee on sterilization that he had "sterilized eleven males with very good results in all cases." The operations were performed at the request of the patients or their relatives. Hotchkiss also reported that he had received letters from two of the sterilized patients reporting that they attributed "the greater part of their well-being and good state of health to the operation that was performed on them."[68] The state's sterilization law was revised in 1927 to limit the criminals targeted by it to only those identified as habitual sexual criminals.[69] Like the Nevada statute, it forbade asexualization by castration or ovariotomy except when performed to remove diseased organs. The act concluded with the claim that "heredity plays a most important part in the transmission of crime, insanity, idiocy and imbecility" and explained that the law was an emergency enactment because "our institutions for degenerates are overcrowded on account of the lack of adequate means of checking the ever-increasing numbers of this class." According to a 1950 article in *North Dakota Bar Briefs* by third-year law student Duane R. Nedrud, "The North Dakota statute has been utilized almost entirely by voluntary methods, and the compulsion provided for it has rarely been used. The explanation probably lies in the danger seen by those in charge of administering the law that a controversy might result in a court decision invalidating the statute."[70] Sixteen years later, Nedrud again demonstrated his support of expanded police powers when he appeared on behalf of the National District Attorney's Association at the U.S. Supreme Court hearing of *Miranda v. Arizona* and argued against Miranda's right to an attorney at any stage of police officer's investigation.[71] North Dakota repealed its compulsory sterilization law in 1965; in total, at least 1,029 people were sterilized under the law, the last 15 of whom were sterilized in 1962.[72]

Authorities who chose to coercively sterilize inmates and patients without securing consent from them or their guardians were rarely challenged with a court trial. Of the more than 63,000 coerced sterilizations in the United States, only a small handful of targeted individuals, their guardians, or occasionally their guardian ad litems challenged the sterilization orders in court. Objections to the operations were typically made based on claims that compulsory sterilization was cruel and unusual punishment, that it violated the constitutional guarantees of due process or equal protection, that it invested judicial power in the hands of the legislature, that it represented an invalid exercise of police power, that it was in essence a bill of attainder, or that it was class legislation. The most widely discussed court case in the American eugenics movement and in the history of coerced sterilization in the United States is *Buck v. Bell*, which focused on the question of whether the state had the authority to require the sterilization of citizens it considered inferior or defective. The next chapter will explore in detail the origins and impacts of the *Buck v. Bell* decision, which powerfully motivated opposition to coerced sterilization. Both before and after the *Buck* decision, however, state courts, including state supreme courts, heard and decided cases that challenged compulsory sterilization laws.

The first court challenge to a compulsory sterilization law came in response to Washington's 1909 law, which called for sterilization of those convicted of "carnal abuse of a female person under the age of ten years, or of rape, or shall be adjudged to be an habitual criminal."[73] The law went unused until 1911, when the Superior Court of King County convicted Peter Feilen of raping a child under the age of ten and sentenced him to life imprisonment at the state penitentiary. In addition, the judge ordered "an operation to be performed upon said Peter Feilen for the prevention of procreation, and the warden . . . is hereby directed to have this order carried into effect . . . by some qualified and capable surgeon by the operation known as vasectomy." Feilen appealed his conviction to the Washington Supreme Court on the grounds that the prosecution had not adequately demonstrated that he had committed the crime and that the sterilization order was cruel and unusual punishment. While the court quickly dismissed his claims about the weakness of the case against him, it spent considerable time considering his challenge to the constitutionality of the law that allowed for his sterilization. In considering Feilen's "brutal, heinous, and revolting" crime, the court asserted that if the legislature saw fit to require it, the penalty of death "might be inflicted without infringement of any constitutional inhibition." Citing claims about the potential social value of compulsory sterilization as well as statements from medical doctors, including Harry Sharp and William Belfield, the court determined that the procedure was no more serious than was the extraction of a tooth. Ultimately, it set precedent in its unanimous decision that ordering an inmate's sterilization as part of the punishment for a crime did not violate the state's constitutional protection against cruel and unusual punishment. For decades, courts around the nation used *Washington v. Feilen* to dismiss claims that compulsory sterilization was cruel and unusual punishment for a crime.[74]

In 1921, the Washington state legislature broadened the range of citizens targeted for coerced sterilization to include the feebleminded, epileptic, and insane, along with morally degenerate persons, sexual perverts, and habitual criminals.[75] The revised law also required the State Department of Health to collect quarterly reports from state institutions on all state wards describing their mental status. The vast majority of Washington's compulsory sterilizations took place during the 1930s; until 1932, the State Department of Health had recorded only 9 sterilizations total, but the number increased to 685 within ten years. During this time, W. N. Keller served as superintendent of the Western State Hospital in Steilacoom, Washington, and he believed that sterilization was a valuable tool in helping treat his mentally ill patients by removing the burden of reproduction from them. There is no evidence in Keller's records that he sterilized patients for punitive or eugenic reasons; rather, sterilization was a therapeutic measure at his institution.[76]

Washington's history of compulsory sterilization ended in 1942 with a judgment by the Washington Supreme Court in the case of Hollis Henderickson, who had been referred for sterilization because he was judged insane by Keller's staff at the Western State Hospital. Henderickson's father filed an appeal with the superior court for Pierce County, claiming that the state's sterilization law was unconstitutional because it violated the due process clause of the Fourteenth Amendment to the U.S. Constitution. The superior court for Pierce County then issued an order that halted all sterilizations performed at state institutions under the 1921 act, and the prosecuting attorney of Pierce County, acting as attorney for the hospital, appealed to the Washington Supreme Court. In a unanimous decision, the court determined that the act did in fact violate citizens' constitutionally protected right to due process because it made no provision for notice to an inmate of the hearing to determine if his or her condition warranted sterilization, nor did it afford the patient the opportunity to appear at the hearing. Moreover, the court determined that the inability of mentally ill or insane citizens to adequately defend themselves in cases in which they lacked a guardian put them at special risk for abuse.[77] Keller and other advocates of sterilization legislation pressed for the passage of a new sterilization law that would overcome the constitutional problems identified in the 1921 law. Forty years later, Washington state legislators passed a "Prevention of Procreation" act in 1961 that stated: "Whenever any person shall be adjudged guilty of carnal abuse of a female person under the age of ten years, or of rape, or shall be adjudged to be an habitual criminal, the court may, in addition to such other punishment or confinement as may be imposed, direct an operation to be performed upon such person, for the prevention of procreation."[78] As of 2006, the law was still on the books, but was not applied by judges.

While the Washington Supreme Court rejected claims that compulsory sterilization was cruel and unusual punishment, the Nevada Supreme Court accepted the argument. Nevada had enacted a compulsory sterilization law in 1911, the same year that *Washington v. Feilen* was decided, but it never coercively sterilized any of its wards.[79] Section 6293 of the Revised Laws of Nevada permitted the sterilization of anyone adjudged guilty of "carnal abuse of a female person under the age of ten years, or of rape, or shall be adjudged to be an habitual criminal," and it banned castration. It was adopted as part of a larger "Crimes and Punishments Bill," passing in both the House and Senate by wide margins.[80] In 1918, seven years after the state's compulsory sterilization law was enacted, an inmate at the Nevada State Penitentiary, Pearley C. Mickle, pleaded guilty to rape and was ordered sterilized. Mickle had epilepsy, a fact that "was accorded considerable weight by the court in pronouncing judgment." He appealed his sterilization to the district court on the grounds that sterilization was cruel and unusual punishment. The court, while acknowledging that the purpose of the statute was to prevent the "transmission of criminal tendencies," decided that because the law did not apply to all convicted offenders, it was in fact a punitive, rather than a eugenic, measure. Ultimately, the court found that, while vasectomy in itself is no more cruel than branding, amputation, slitting of a tongue, or the cutting off of an ear, when used as punishment vasectomy "is ignominious and degrading, and in that sense cruel." Mickle was spared the punishment of vasectomy, and the state of Nevada, while being included on the list of states that passed a compulsory sterilization law, never coercively sterilized one of its citizens.[81]

Many of the challenges to compulsory sterilization laws focused on the laws' violation of the constitutional guarantee of equal protection, which is derived from the Fourteenth Amendment and requires that laws not disproportionately impact particular classes of citizens. For example, Iowa's 1911 sterilization law focused on prisoners convicted of crimes that suggested they were moral or sexual perverts and included in its purview "criminals, rapists, idiots, feebleminded, imbeciles, lunatics, drunkards, drug fiends, epileptics, syphilitics." It created a separate set of penalties for men who forced women into prostitution that prescribed one to ten years in the state penitentiary.[82] The first compulsory sterilization in the state was performed a year later on a man convicted of sodomy and diagnosed as a sexual pervert with dementia praecox.[83] In 1913, legislators passed a second compulsory sterilization law that added "moral and sexual perverts, and diseased and degenerate persons" as targets for compulsory sterilization. The second law also created a way for people with syphilis and epilepsy to apply to district judges for voluntary sterilization, which would allow them to get around the state's marriage laws and legally marry. In 1914, the U.S. District Court of the Eastern District of Iowa heard *Davis v. Berry et al.*, in which three defendants, each of whom had been convicted of a crime and ordered sterilized, alleged that the law violated the U.S. Constitution's guarantees of equal protection and due process and the prohibition of cruel and unusual punishment, as well as the state constitution's provision for the right to pursue and obtain safety and happiness. The court concluded that the law, because it "automatically decides the question and nothing remains for the prison physician to do but to execute that which is already of record," was essentially a bill of attainder. As such, it declared Iowa's original sterilization law unconstitutional and therefore void.[84] The U.S. Supreme Court upheld the decision when it declared the state's appeal moot in a brief decision written by Oliver Wendell Holmes Jr.[85] In 1915, Iowa state legislators enacted a third compulsory sterilization law, which they revised in 1929 and again in 1946 with provisions intended to overcome the earlier laws' shortcomings in providing for due process. Like earlier laws, the 1915 law cast a broad net and included the feebleminded, insane, syphilitic, habitual criminals, moral degenerates, and sexual perverts.[86] The 1946 law created the State Board of Eugenics, which was required to meet quarterly and submit the names of all persons living in the state it knew to be "feebleminded, insane, syphilitic, habitual criminals, moral degenerates, or sexual perverts and who are a menace to society." A provision in the law also allowed the inclusion of anyone who in the judgment of the board would produce a child who "would probably become a ward of the state." The board interpreted this section of the law to be an environmental provision, allowing for the sterilization of individuals damaged by their environment or who provided a damaging environment to their children.[87] As was the case in North Dakota, the compulsory sterilization laws adopted in Iowa after 1915 were not challenged in court because before they operated, doctors generally obtained consent from the patient or someone who could legally consent for the patient.[88] Coerced sterilizations continued in Iowa through 1963, when the last thirty citizens were sterilized. In total, 1,910 people were sterilized under Iowa's five sterilization laws, with half of the sterilizations taking place after 1948 and with the largest single-year total number of sterilizations being 178 in 1951.[89]

In 1911, state legislators in New Jersey passed an act, signed into law by then-governor and future-president Woodrow Wilson, that established a board of experts to examine the "mental and physical condition of the feeble-minded, epileptic, certain criminals and other defective inmates confined in the several reformatories, charitable and penal institutions in the counties and state." The "certain criminals" were those who had been convicted of rape. Like the nation's first compulsory sterilization law passed by Indiana's legislators, New Jersey's sterilization law stated that "heredity plays a most important part in the transmission of feeble-mindedness, epilepsy, criminal tendencies, and other defects." It required the superintendents of the institutions that housed targeted individuals to apply to the board for authorization to sterilize their wards.[90] Two years after the law was enacted, a lawyer for Alice Smith, an inmate of the New Jersey State Village for Epileptics, petitioned to stop David Weeks, the village's chief physician, from sterilizing her. Weeks and the Board of Examiners of Feeble-Minded, Epileptics, Criminals, and Other Defectives had declared Smith an epileptic, and under the existing statute she was to be sterilized.

In *Smith v. Board of Examiners*, the New Jersey Supreme Court deliberated on the question of whether the state's compulsory sterilization statute was an invalid exercise of police power and declared it unconstitutional, making New Jersey the first state to have its sterilization law invalidated. The majority opinion described both the existing law and the salpingectomy operation, as well as the category of epileptic, into which Alice Smith was classified. The court found that while the existing law addressed only the feebleminded, epileptic, and certain criminals, if it was allowed to stand it could justifiably be expanded to include other groups. Whose elimination, the court asked, "in the judgment of the legislature, [would] be a distinct benefit to society[?] If the enforced sterilized of this class be a legitimate exercise of government power, a wide field of legislative activity and duty is thrown open to which it would be difficult to assign a legal limit." Those with specific diseases, such as pulmonary consumption or communicable syphilis, could be singled out as needing to be specially treated for the protection of society. At that point, racial differences "might afford a basis for such an opinion on communities where that question is unfortunately a permanent and paramount issue." Moreover, the fact that the law pertained only to those in state institutions made it "singularly narrow" and therefore not in keeping with the broad purpose of the statute. Finally, the court concluded, "the suggestion that . . . the scheme of the statute were to turn the sterilized inmates of such public institutes loose upon the community and thereby to effect of saving of expense to the public, is not deserving of serious consideration" because the "palpable inhumanity and immorality of such a scheme forbids us to impute it to an enlightened legislature."[91]

New York's 1912 sterilization law focused on the state's "feeble-minded, epileptic, criminals and other defectives" as determined by a board of examiners consisting of a surgeon, a neurologist, and a medical practitioner. If a majority of the board determined that "such person would produce children with an inherited tendency to crime, insanity, feeble-mindedness, idiocy or imbecility, and there is no probability that the condition of any such person will improve," the board could authorize sterilization. The criminals subjected to the law were those convicted of rape or of a succession of offenses such that "as in the opinion of the board shall be deemed in the criminal examined to be sufficient evidence of confirmed criminal tendencies."[92] In 1918, Frank Osborn, a twenty-two-year-old man who was "in a class known as feeble-minded" and a longtime resident of the Rome Custodial Asylum, challenged the law that allowed the Board of Examiners of Feeble-Minded Criminals and Other Defectives to order him sterilized.[93] His case originated from an inquiry into the constitutionality of the state's compulsory sterilization law by the state's attorney general, and the trial brought to the courtroom of the New York Court of Appeals a series of experts on eugenics, mental health, and heredity who made clear statements about their beliefs regarding compulsory sterilization, most stating that they opposed it.[94] Dr. Lemon Thomason, one of the three members of the board of examiners, testified that he had performed a superficial examination of Osborn and his family and that the operation would bring no benefit to the patient, nor would it weaken in him "the tendency of the rapist." Another board member stated that a vasectomy would, in his opinion, do nothing to help the patient. Castration, though, would be effective in curbing his tendencies to rape. The superintendent, Dr. Bernstein, testified about his knowledge of heredity: "We are taught that the dominant traits appear in three-quarters of the offspring and recessive traits appear in one-quarter, when the parentage is mixed as regards traits; that it is only in cases of feeble-mindedness of both parents that you would look generally for an increase of feeble-mindedness among offspring." Despite his beliefs about the Mendelian inheritance of feeblemindedness, Bernstein did not support sterilizing his feebleminded charges and claimed that enforcement of the state's sterilization law would "create a class of people by themselves who would feel that they were so different from normal humanity that they would go back to promiscuous sexual relations and that there would be known places where these people were harbored and there they would tend to collect." Regardless of whether or not he was sterilized, Osborn would have to be supervised constantly because "society needs protection from the raping of little girls and the frightening of them just as much as it wants protection from a future generation of dependents and delinquents."[95]

Charles Davenport also testified at the trial, stating that he generally agreed with the statements made by Dr. Sharp of Indianapolis in an article entitled "Vasectomy as a Means of Preventing Procreation in Defectives." Defective persons, Davenport claimed, "are not necessarily to become a public charge," because included in the class labeled defective are "the most gifted as well as the most vicious, weakest and ordinarily the most unhappy of mankind," and listed Thomas Chatterton, Oliver Goldsmith, Samuel Coleridge, and Charles Lamb among them. He also explained that the New York sterilization law grew "out of studies and efforts of those who are interested in the subject of eugenics" and was based on the "laws of heredity; with improving through better breeding." Davenport mentioned the studies of the Jukes and the Nams in support of his claim that "there is to be found much of good in the most degenerate families known in our land." Ultimately, Davenport "testified that he has not advocated the operation of vasectomy, and that in his opinion segregation of the sexes would be better." Bleecker Van Wagenen, chairman of the American Breeders Association Committee on Sterilization, joined Davenport in testifying that voluntary sterilization was ultimately beneficial, but "when such operations have been done against the will of the patient the psychic effects have been bad." Finally, Walter Fernald, superintendent of the School for the Feeble-Minded in Massachusetts, testified "that he had never seen an authorized medical statement based upon actual facts which would justify claims made for the results in Indiana, where such a law is in operation." Moreover, he believed that sterilizing the feebleminded would result in increased "illicit intercourse" and that ultimately the effect of the operation would be to exchange the "burden of feeble-mindedness for the burden of sex immorality or sex diseases and of insanity resulting from that condition." He feared that widespread sterilization of the feebleminded would result in prejudice by "many right-thinking persons who are interested in those who are afflicted against institutions."[96]

The court concluded that the board of examiners knew "very little about the subject. They have given it no particular study. They are not, in the opinion of the court, justified in the determination which they have reached." Reviewing the board's work, the court reversed the decision to sterilize Osborne. The unanimous decision then turned to address the question of the constitutionality of New York's 1912 compulsory sterilization law. Osborn's counsel had claimed that the law violated the U.S. Constitution because it was a bill of attainder, that it deprived a citizen of a trial by jury and of the privileges or immunities of a citizen, that it compelled citizens to be witnesses against themselves, that it violated due process, that it permitted cruel and unusual punishment, and finally that it denied citizens equal protection of law. Relying on the 1913 New Jersey Supreme Court decision in *Smith v. Board of Examiners of Feeble-Minded*, the New York court declared that the law allowed an unjustifiably broad exercise of police power that was "almost inhuman in its nature."[97] The court declared the New York compulsory sterilization law unconstitutional after forty-two inmates or patients had been sterilized under it.[98] Two years later, in 1920, the New York legislature unanimously repealed its sterilization law.[99]

Michigan's compulsory sterilization law was eventually declared an unconstitutional violation of patients' and prisoners' rights because it was, according to the state supreme court, class legislation. The state's 1913 law authorized "the sterilization of mentally defective persons maintained wholly or in part by public expense in public institutions in this State, and to provide for the unauthorized use of the operations provided for." It targeted residents of publicly supported institutions and limited sterilization operations to vasectomies and salpingectomies, or by the "surgical operation which is least dangerous to life and will best accomplish the purpose."[100] Ten years later, in 1923, Michigan legislators enacted a law that applied to "any mentally defective person, including all feebleminded, insane, epileptic, idiots, imbeciles, moral degenerates and sexual perverts, who would be likely to procreate children with a tendency to mental defectiveness."[101] This law was challenged in 1925 in *Smith v. Wayne Probate Judge*, and the court concluded that the law did not violate the defendant's rights to freedom from cruel and unusual punishment and accepted that its targeting of feebleminded citizens did not constitute class legislation. However, the court agreed with the defendant's claim that compulsory sterilization of those citizens who were unable to financially support the children that they produced was in fact class legislation and therefore made the law problematic.[102] In response to the challenges presented by *Smith v. Wayne Probate Judge*, in 1929 Michigan state legislators repealed the 1923 law and replaced it with an "act to prevent the procreation of feeble-minded, insane and epileptic persons, moral degenerates, and sexual perverts; to authorize and provide for the sterilization of such persons and payment of the expenses thereof." It also allowed for a broad number of petitioners, any one of whom could sue to prevent the sterilization.[103] In 1943, the legislature passed an act to provide for the funding of sterilizations, allocating up to $50 for each sterilization when performed by a surgeon not already employed by a state institution.[104] As was the case in several other states, Michigan's legislators found ways to get around the state court's constitutional challenges to coerced sterilization.[105]

In Alabama, as in Michigan, state supreme court justices believed that compulsory sterilization laws that targeted state wards were class legislation and were therefore unconstitutional. In 1935, Alabama governor Bibb Graves wrote to the justices of the Alabama Supreme Court requesting their opinion on the constitutionality of the state's compulsory sterilization laws. Graves wanted to know if the law was a valid exercise by the legislature of police power, whether it was based upon a reasonable classification because it failed to include individuals not housed in state institutions, and whether it violated constitutional prohibitions against cruel and unusual punishment. The justices found no difficulties with the state's compulsory sterilization law, except with respect to its targeting of only those defectives housed in state institutions. This, they believed, violated the

Fourteenth Amendment's guarantee of due process, because it deprived the patients of some aspect of themselves without a hearing before a duly constituted tribunal or board. Even recognizing the precedent established in *Buck v. Bell*, they stated that the lack of a hearing made Alabama's compulsory sterilization law unconstitutional.[106] Nonetheless, the law apparently operated unchanged until 1974, when a federal court intervened to establish "adequate standards and procedural safeguards to insure that all future sterilization be performed only where the full panoply of positional protections has been accorded to the individuals involved." These included ensuring that the sterilization was in the best interest of the patient, the patient was at least twenty-one years old and had given appropriate consent, and the order was vetted by an appropriate review committee.[107]

Perhaps the most significant blow to compulsory sterilization laws came in 1942 in response to Oklahoma's 1931 law, which specifically targeted habitual criminals, defined as any person convicted of three separate felonies.[108] The Oklahoma Supreme Court had upheld the constitutionality of the law in 1933, declaring that it was not a violation of the state constitution's provision for due process, nor was it cruel and unusual punishment or a violation of a citizen's right to life, liberty, and happiness.[109] The law was revised twice, once in 1933 and again in 1935, to increase the due process protections by requiring a formal appeals process.[110] At the same time, the number of felony convictions required to initiate sterilization proceedings decreased from three to two, while the nature of the felonies that made inmates eligible for sterilization was narrowed to only those felonies that involved "moral turpitude." Eliminating crimes such as embezzlement from the list of felonies that made one eligible for sterilization would ultimately lead to the demise of the state's compulsory sterilization law. In 1941, the Oklahoma Supreme Court again upheld the state's compulsory sterilization law in *Skinner v. State*. Skinner was convicted of stealing chickens in 1926, then in 1929 and 1935 of the crime of robbery with firearms. The three felonies made him eligible for sterilization under Oklahoma's compulsory sterilization law. The state supreme court upheld the law based on the fact that "statistics, scientific works, and information from which it found as a fact that habitual criminals are more likely than not to beget children of like criminal tendencies who will probably become a burden on society."[111] The following year the U.S. Supreme Court heard the case and unanimously overturned the lower court's decision, with three justices offering separate opinions in support of their decision. The majority opinion declared the law unconstitutional because it violated citizens' Fourteenth Amendment rights to equal protection because it did not treat all repeat felons equally: "When the law lays an unequal hand on those who have committed intrinsically the same quality of offense and sterilizes one and not the other, it has made as invidious a discrimination as if it had selected a particular race or nationality for oppressive treatment." In a pithy statement that rivaled Justice Holmes's infamous claims in *Buck v. Bell*, Justice William O. Douglas concluded, "Embezzlers are forever free. Those who steal or take in other ways are not."[112] With the exception of seventy-two reported operations in 1952, compulsory sterilizations in Oklahoma ended in 1942 with *Skinner v. Oklahoma.*[113] It took another four decades for the many states that adopted compulsory sterilization laws to finally stop coercively sterilizing their citizens.

While there were a number of challenges to compulsory sterilization and several courts decided that some aspects of a state's sterilization law were unconstitutional, there were also several state's whose courts gave blanket approval for the coerced sterilization of unfit citizens. Take, for example, the situation in Idaho. Legislators there passed the state's first compulsory sterilization bill in 1919, which was vetoed by Governor D. W. Davis. The bill targeted the "feeble-minded, insane, epileptic, moral degenerates and sexual perverts" who were inmates of institutions maintained at public expense.[114] In his veto message, Davis explained that he did not believe that the bill would have accomplished its aim of preventing procreation by degenerate citizens. "It does not apply to all persons in such classes, but only to those confined in public institutions—the persons in fact who by reason of such confinement are the least menace to society." Such laws, he explained, had been declared unconstitutional in other states because they applied only to a certain class of citizens, and he concluded, "The scientific premises upon which these laws are based are still too much in the realm of controversy and the results of the legislation still too experimental to justify the proposed law as wise legislation for this state." The legislature did not attempt to overturn Davis's veto.[115] Six years later, in 1925, legislators again approved a compulsory sterilization law, which was signed into law and revised four years later.[116] Eugenic sterilization advocates lauded the Idaho law because it applied not only to inmates of state institutions but to all people living in the state. Thus the state board of eugenics had the legal authority to order the sterilization of anyone it believed had an inherited tendency to feeblemindedness or would probably become either a social menace or a ward of the state. The *Eugenical News* reported in 1931, "No eugenicist interested in the constitutional aspect of eugenical sterilization could demand a better legal situation than that which exists in Idaho."[117] In 1931, the Idaho Supreme Court heard *State v. Troutman*, in which the guardian of Albert Troutman, who authorities had declared was "afflicted with congenital feeblemindedness and recommended sterilization by vasectomy," had sued to stop the operation on the grounds that it represented an unreasonable exercise of police power, it was cruel and unusual punishment, it was punitive, it delegated judicial powers to the executive, and finally it violated due process and equal protection. The court refused every one of Troutman's claims and decided that the operation was "neither seriously dangerous nor painful, nor at all injurious to the physical health or happiness of the individual upon whom it is performed."[118] It concluded, "The law in behalf of the general welfare demands it should be applied in this case," and Troutman became one of the thirty-eight people coercively sterilized by the state of Idaho.[119] Coerced sterilizations appear to have ended there in 1963, and Idaho's legislature repealed the state's compulsory sterilization law in 1972.[120] Five years later, legislators decisively limited the state's authority to sterilize its wards, stating, "The legislature of the state of Idaho acknowledges that sterilization procedures are highly intrusive, generally irreversible and represent potentially permanent and highly significant consequences for individuals incapable of giving informed consent. The legislature recognizes that certain legal safeguards are required to prevent indiscriminate and unnecessary sterilization of such individuals, and to assure equal access to desired medical procedures for all Idaho citizens."[121]

The Nebraska Supreme Court gave blanket approval for compulsory sterilization in 1968 when it decided the case of *The State of Nebraska v. Gloria Cavitt*. Cavitt had been committed to the Beatrice State Home and been ordered sterilized prior to being paroled. Her guardian sued the state on the grounds that Cavitt's sterilization order was unconstitutional because it was an unreasonable exercise of police power, because it denied her right to equal protection, because the use of the term "mentally deficient" in the law was too vague and indefinite, because it delegated quasi-judicial powers to a board of examiners and was an unlawful delegation of legislative power, because it did not afford constitutional procedural due process, and finally because the operations were inhumane, unreasonable, and oppressive punishments for crime. The Nebraska Supreme Court denied every one of the challenges and concluded that the state's compulsory sterilization laws were "in all respects constitutional and enforceable."[122] The following year, the state legislature repealed its sterilization law.[123]

In 1972, the Court of Appeals of Oregon likewise gave comprehensive approval for compulsory sterilization when it decided the case of *Cook v. State*. The case resulted from an order by the State Board of Social Protection for the sterilization of Nancy Rae Cook, "a 17-year-old girl with a history of severe emotional disturbance" and a ward of the state for the previous four years.[124] The Board of Social Protection had filed a petition for her sterilization, based on the 1967 sterilization law, after Cook had "engaged in a series of indiscriminate and impulsive sexual involvements while she was in the hospital." Cook's lawyer sued to stop the sterilization, claiming that the act violated her right to equal protection under both the state and federal constitutions. In determining the case, the court concluded, "The state's concern for the welfare of its citizenry extends to future generations and when there is overwhelming evidence, as there is here, that a potential parent will be unable to provide a proper environment for a child because of his own mental illness or mental retardation, the state has sufficient interest to order sterilization."[125]

Four years later, in 1976, the North Carolina Supreme Court upheld the constitutionality of the state's sterilization law against the claims that it was a violation of due process, an invalid exercise of police power, and a violation of equal protection, and that it represented cruel and unusual punishment. *In re Joseph Lee Moore*, a case in which the director of a county department of social services petitioned for the sterilization of a mentally deficient man, the court supported the right of the state to sterilize incompetent citizens. Citing both *Roe v. Wade* and *Buck v. Bell*, the majority decision claimed, "The Right to procreate is not absolute but is vulnerable to a certain degree of state regulation," and "the interest of the unborn child is sufficient to warrant sterilization of a retarded individual."[126] In 1976, the same year the state supreme court decided the case, North Carolina legislators passed an act that allowed for the sterilization of people with mental disabilities when the person "would be likely, unless sterilized, to procreate a child or children who would have a tendency to serious physical, mental, or nervous disease or deficiency which is not likely to materially improve."[127] The law was repealed in 1987.[128] Ultimately, North Carolina's sterilization laws led to the coerced sterilization of more than 7,600 people.[129]

The Vas Deferens Is Not Enough

Reproductive capacities were not the only targets of doctors' scalpels and legislatures' acts. In several states, men in mental health hospitals and prisons were not just sterilized, they were mutilated. Doctors in Oregon, Kansas, Michigan, Indiana, Pennsylvania, and perhaps other states castrated hundreds of men in their charge beginning in the 1890s and continuing until at least the 1930s. Over the course of these four decades, justifications for castration shifted from punitive and eugenic to therapeutic and prophylactic. The first castrations were done in the years before the advent of the vasectomy and were justified on the same bases as were the thousands of hysterectomies that were done on so-called hysterical women. From the 1880s through the second decade of the twentieth century, the men and women targeted for castration were generally considered addicted to masturbation. Justifications for the operations generally focused on their potential to relieve victims of impulses they simply could not resist, which lightened the loads of both the patients and the health care professionals charged with taking care of them.

By the second decade of the twentieth century, an increasing number of mental health patients and convicted felons were castrated because they had either raped children or been convicted of committing homosexual acts, and the use of castration operations shifted from the therapeutic to the punitive. Later advocates of castration operations shared with their predecessors the expectation that the procedures would relieve their patients' uncontrollable urges. The difference, though, rested in the role of legislative requirements for the sterilization of anyone deemed feebleminded or convicted of a particular felony or a particular number of felonies. The legal obligation for mental health care providers and prison wardens to sterilize their patients, an obligation for which many of them had campaigned and many more of them happily complied, added a punitive element to the surgeries that had not existed prior to 1907.

Turning to Oregon as an example, we see how the choice to castrate or merely sterilize men was made. Between 1912 and 1983, the state's physicians coercively sterilized over 2,200 patients in mental health asylums and inmates in state prisons. Of these, at least 184 men were castrated, and in examining their case files, one finds that most of them had been convicted of a sex crime, usually either rape, child molestation, or some form of crime associated with homosexual activity.[130] Coming on the heels of a same-sex vice scandal in Portland, the state's sterilization laws were motivated at least in part by concerns about homosexuality and other so-called sexual and moral perversions. While vasectomies were widely performed in the state's mental health and penal institutions, over two-thirds of the men sterilized at the State Hospital in Salem were castrated. In previous studies of involuntary sterilization, historians have argued that "castration was too brutal to provide a socially acceptable solution to curbing the fecundity of the feebleminded."[131] That was obviously not the case in Oregon, where officials were apparently interested in doing more than simply removing the traits of rapists, homosexuals, and child molesters from future generations; authorities wanted to unsex them.[132] Once the vasectomy was invented and in use, as it was in Oregon, a doctor would have employed it, rather than castration, if all he or she wanted to do was sterilize the patient. Castrating a patient provided no more eugenic value than did a vasectomy and was considerably more dangerous and invasive an operation, so one can only assume that the physicians believed that castrations had additional punitive or therapeutic benefits.

Challenges to compulsory sterilization laws in state courts demonstrated that a carefully crafted bill—one that included opportunities for appeals, was not punitive in nature, and applied equally to patients in state and private institutions—usually overcame constitutional challenges. The ongoing question with the laws was whether the state had a compelling interest and the police power to limit the reproductive powers of certain citizens. In 1927, *Buck v. Bell* decisively answered this question in the affirmative, and in so doing it empowered states to pass compulsory sterilization laws and practice coerced sterilization. It also gave the unorganized, largely ineffectively collection of sterilization opponents a target around which they could unite.

CHAPTER 4

Buck v. Bell and the First Organized Resistance to Coerced Sterilization

Before the late 1920s, the only organized resistance to compulsory sterilization laws came from local or regional antisterilization groups. Take, for example, Lora Little's Anti-Sterilization League, which organized in 1913 to oppose Oregon's compulsory sterilization law. Little's opposition to sterilization was part of her broader animosity toward the medical profession motivated by the death of her seven-year-old son. She believed that her son had died from a reaction to a smallpox vaccination, and she equated compulsory sterilization and compulsory vaccination. She "considered doctors to be little more than power- and profit-hungry oppressors who, operating with faulty ideas, only made people sicker." Little led the drive for a referendum on Oregon's sterilization law, which ultimately overturned the law and prevented the implementation of coercive sterilization in the state for several years.[1]

Organized resistance to compulsory sterilization laws was limited to local or regional opposition because of the nature of the laws themselves. Issues of public health and oversight of both medicine and education—the aspects of government for which compulsory sterilization was relevant—were under the purview of state, rather the federal, governments. The sterilization laws themselves and the actual practice of coerced sterilization, both in terms of its targets and the frequency with which it was applied, varied enough to make the basis for opposition to them differ considerably from state to state. The disconnected collection of sterilization opponents needed something that would unite them in opposition to compulsory sterilization laws, which finally came with the decision in *Buck v. Bell*, the U.S. Supreme Court's affirmation of states' abilities to enact compulsory sterilization laws. *Buck v. Bell* provided a rallying cry, especially among Catholics, and it galvanized opposition to compulsory sterilization, which was led by the only nationwide organization to oppose compulsory sterilization, the Roman Catholic Church.[2]

Voices of Opposition before *Buck*

As ardent as were the supporters of sterilization for the prevention of crime in the late nineteenth century, there were some equally fervent opponents to the procedures, and they occasionally published letters and articles in medical journals. They were not, however, particularly effective. Among the first opponents to publish in the professional literature was J. W. Lockhart, a medical doctor in St. John, Washington, who in 1895 wrote in the *St. Louis Courier of Medicine* describing his opposition to "bodily mutilation." Society could be made safe not through sterilization, but only through "the education of its component parts above a certain moral standard, demanded and maintained by public sentiment." Lockhart ignored eugenic arguments and, applying the same sort of biological analogies that would be common a decade later, explained that it was "useless to lop off the withered branches while the worm is eating away the root of the tree." Time spent inventing new punishments for crime would be much better spent developing the institutions that educated citizens and prevented crime. Ultimately, he argued, "the medical profession should lead in this great reform."[3]

Many of the authors who opposed some aspect of coerced sterilization believed that the approach still had some merit, and they would often criticize certain uses of it while praising others. Take, for example, the physician A. C. Corr, who wrote "Emasculation and Ovariotomy as a Penalty for Crime and as a Reformatory Agency" in 1895. Like Lockhart, Corr generally argued against the use of sexual surgeries as a punishment for the commission of crime, emphasizing that "the criminal tendency is a mental complex, a moral imbecility, congenital or hereditary, but in either respect largely the result of environment and synergistic influences." But even though he generally opposed laws that would bring compulsory sterilization to bear on convicted criminals, he agreed that emasculation should "be applied as a punishment—precautionary—for rape, because one who has such a perverted impulse should, for the safety of society, be rendered unable to commit the deed involved."[4] This sort of limited rejection of compulsory sterilization, arguing against it for the vast majority of targeted individuals but holding out some truly incorrigible group that deserved the scalpel, was common in the literature into the second half of the twentieth century. Even the most aggressive opponents of coerced sterilization often set aside some particularly problematic group for the procedures.

Some medical practitioners saw in the widespread use of the vasectomy to control unfit citizens' reproduction the danger of social upheaval by sterile, but potent and degenerate, people. In a 1910 article in the *New England Medical Monthly*, Francis Barnes argued that, while a patient who received the operation "becomes of a more sunny disposition, brighter of intellect, ceases excessive masturbation and advises his fellows to submit to the operation for their own good," some criminals may seek the operation because it enlarges their "opportunities for illicit sexual intercourse by removing the danger of resulting pregnancy."

Barnes clearly appreciated the optimistic claims about compulsory sterilization's supposed abilities to check the "growth of degeneracy," but worried that the laws allowed the states "the right to mutilate those of its citizens whom a committee of experts may consider unfit to procreate offspring." He concluded that while the "actual number of real defectives is certainly large" and financially burdensome, the "graver objections to be urged against sterilization" outweighed any potential benefits promised by zealous promoters of compulsory sterilization laws.[5]

Abraham Myerson, a professor of neurology at Tufts College Medical School, was among the first American professionals to voice effective and prolonged opposition to compulsory sterilization laws in the United States. At the 1924 meeting of the Boston Society of Psychiatry and Neurology, he presented his first attack on coerced sterilization. The terms *insanity* and *feeblemindedness*, he asserted, imply that the broad range of conditions they denote are unit characters, when in fact they are "pernicious in that they exercise an influence on the mind, which vitiates most of the work done on the inheritance of mental disease." The fact of the matter, Myerson argued, was that there "was not unity in feeblemindedness." He singled out the work done by Goddard and Davenport as "extremely faulty" and "unscientific" because they rubberstamped individuals as feebleminded from a mere glance at a court record. Myerson accepted the inheritance of mental diseases, but concluded that the view currently in vogue led to the "adoption of a policy that does not aim either at investigation of causes or at therapeutics."[6] The discussion that followed his presentation almost universally supported his claims. It demonstrated an emerging turf battle between biologists and the neurologists and psychiatrists who had assembled at the meeting; again and again they made clear the need to take back from the biologists discussions about mental illness and feeblemindedness. In the search for the Mendelian law among their feebleminded research subjects, the psychologists and neurologists claimed that biologists had grossly oversimplified the root cause of diminished intelligence. As Walter Fernald asserted, "It is a far cry from a study of white and red peas to the study of human intelligence. The study of human intelligence implies a complex combination of factors."[7]

The following year, Myerson published *The Inheritance of Mental Diseases*, in which, as he had done in his talk the previous year, he singled out Goddard's 1914 *Feeblemindedness* along with several of Davenport's works for special criticism. Davenport, he charged, seemed "determined to find in the mental diseases a Mendelian significance," which he did by postulating "in advance conclusions which he is able to verify." On the subject of the inheritance of epilepsy, a problem that Myerson explained had several potential causes, "Davenport and his followers have been dogmatic offenders against logic and science—they have collected data in a thoroughly unscientific way, they have unified utterly diverse conditions into one 'neuropathic defect due to lack of a unit determiner' which . . . is an arbitrary conclusion decided upon, apparently, beforehand."[8] Drawing from David Heron's critiques of Davenport's work, Myerson offered some of the same claims against him that Pearson and the biometricians had made.[9]

A decade after his initial criticism of the biologists' claims about the inheritance of mental defects, Myerson began speaking out against compulsory sterilization laws. In 1935, he published "A Critique of Proposed 'Ideal' Sterilization Legislation" in the *Archives of Neurology and Psychiatry*. As with many other opponents to coerced sterilization, Myerson simultaneously attacked compulsory sterilization laws while he upheld the validity of sterilizing some people. "Personal researches concerning the hereditary factor in mental disease and mental defects," he wrote, "have led to the logical conclusion that from the biologic standpoint there are persons who should be sterilized in order that their particular type of mental disease or defect shall not be transmitted to succeeding generations." A limited program of sterilization, Myerson believed, "is eugenically sound." While he was "in sympathy with limited sterilization laws," he attacked "as extravagant and as approaching mania a proposed sterilization law set up as the ideal by certain important eugenicists." Laughlin had proposed this law several years earlier in *The Legal Status of Eugenical Sterilization*, and it was aimed at preventing the "procreation of persons socially inadequate from defective inheritance."[10] Social inadequacy, as a condition for sterilization, struck Myerson as an even more drastic and less defensible extension of Davenport's notion of a unit character for feeblemindedness. "Who shall say who is a useful person?" he asked. He suggested the law could be used to coercively sterilize Communists, capitalists, Jews, artists, any innovator, and "those restless and reckless persons who fail because they attempt too much, but who are the ferment by which the mass is lifted."[11] Myerson continued his attack on certain aspects of coerced sterilization throughout the 1930s and 1940s, and had his most influence when he chaired the American Neurological Association's Committee for the Investigation of Sterilization in the mid-1930s.

By and large, early opponents to coerced sterilization were lone voices in their professions, and they were inconsistent in attacking some aspects of the compulsory sterilization laws while supporting others. Before the late 1920s, it was rare to find an author who unequivocally rejected the idea of coercive sterilizations as therapy, for punishment, to make it easier to house patients in mental health institutions, or to improve the nation's overall genetic quality. That changed once opponents to sterilization had a single, nationwide target at which to direct their attention, which finally came with the 1927 decision in the *Buck v. Bell* case.

The impact of the *Buck v. Bell* decision in galvanizing opposition to compulsory sterilization laws is very much similar to the effects Michael Klarman describes the 1954 *Brown v. Board of Education* decision had in inspiring massive southern resistance to the civil rights movement. Klarman argues that *Brown*'s direct effect on school desegregation was limited and maintains that "scholars may have exaggerated the extent to which the Supreme Court's school desegregation ruling provided critical inspiration to the civil rights movement." Focusing on the "backlash" to *Brown*, he concludes that the case "crystallized southern resistance to racial change, which—from at least the time of Harry S. Truman's civil rights proposals in 1948—had been scattered and episodic."[12] In much the same way the backlash to *Brown* involved the coming together of previously unorganized opposition to racial integration, *Buck* united the opponents to compulsory sterilization. The impact of the *Buck* decision in motivating opposition to compulsory sterilization is recognized even by the few modern-day advocates of eugenics, such as Richard Lynn. For example, in his 2001 *Eugenics: A Reassessment*, just before launching into an attack on Stephan Jay Gould's analysis of *Buck v. Bell*, Lynn argues, "The case of Carrie Buck became a focus for the opposition to compulsory sterilization of the mentally retarded, which gathered momentum in the second half of the twentieth century."[13]

"Three Generations of Imbeciles"

Buck v. Bell, specifically Justice Oliver Wendell Holmes Jr.'s majority opinion, is infamous in the American eugenics movement and in the history of coercive sterilization in the United States. The case came before the U.S. Supreme Court during its October 1926 term from the Supreme Court of Appeals of Virginia. It involved Virginia's 1924 compulsory sterilization law and the family of Carrie Buck, an eighteen-year-old resident of the Lynchburg State Colony for Epileptics and Feebleminded. Buck had been committed to the colony by the family that adopted her shortly after she had become pregnant; it was later learned that her pregnancy was the result of her rape by the nephew of her adoptive parents, and her commitment was part of the family's attempt to cover up the incident. As directed by Virginia's compulsory sterilization law, the colony's superintendent, Dr. Albert Sidney Priddy, examined her and determined that she had the mental age of a nine-year-old, so he filed a petition for her sterilization. Several supporters of Virginia's compulsory sterilization law, including Priddy, asked Buck's guardian to challenge the sterilization order so that the law on which it was based could be sanctified by the courts. The case moved through the circuit court, to the state's supreme court, and ultimately to the U.S. Supreme Court. Because the case was a friendly one in which the goal was to merely confirm the legality of the state's compulsory sterilization law, Buck's lawyers did not challenge her categorization as defective.[14]

At trial, J. H. Bell, the newly appointed superintendent of the colony, testified that Carrie Buck was the feebleminded, illegitimate child of a likewise feebleminded woman. Like her mother, Carrie had given birth to an illegitimate child who gave "evidences of defective mentality." Harry Laughlin testified by deposition, providing an analysis of the "hereditary nature of Carrie Buck" in which he stated, "All this is a typical picture of a low grade moron. . . . The family history record and the individual case histories, if true, demonstrate the hereditary nature of the feeble-mindedness and moral delinquency described in Carrie Buck. She is therefore a potential parent of socially inadequate or defective offspring." The court also heard testimony from other state mental health officials and eugenicists, all of whom argued that the Bucks' feeblemindedness was hereditary.[15]

Figure 4.1. Carrie Buck and her mother, Emma. Photo from the Arthur Estabrook Papers, M.E. Grenander Department of Special Collections and Archives, University of Albany, SUNY

After hearing Carrie Buck's case, on May 2, 1927, the Supreme Court issued an eight-to-one decision stating that she, as well as her daughter and mother, were feebleminded and therefore prone to promiscuity. Citing the many state decisions in support of compulsory sterilization laws, the Court concluded that Virginia's law did not impose cruel and unusual punishment on those identified as feebleminded, and it afforded necessary due process of law in ordering their sterilizations. The Court also rejected claims that the law was an invalid exercise of police power, finding that the exercise of power allowed by compulsory sterilization laws was no greater than that exercised in compulsory vaccination laws, which were within the legitimate power of the state.

In his majority opinion, which one biographer called his "strongest, most pungent" opinion, Holmes described Carrie Buck as "a feeble minded white woman who was committed to the State Colony."[16] The daughter of a feebleminded mother and the mother of a feebleminded child, she was duly ordered sterilized under Virginia's 1924 compulsory sterilization statute. Holmes discussed the purpose of the sterilization law as well as the procedure for ordering a feebleminded state ward sterilized. There were no procedural errors claimed, but rather a substantive challenge to the law on the grounds that the law inappropriately extended the powers of the state. Holmes rejected the claim, asserting, "We have seen more than once that the public welfare may call upon the best citizens for their lives. It would be strange if it could not call upon those who already sap the strength of the State for these lesser sacrifices, often not felt to be such by those concerned, in order to prevent our being swamped with incompetence." Citing *Jacobsen v. Massachusetts*, Holmes argued that since the Court had already allowed for compulsory vaccination laws, it was justified in supporting compulsory sterilization laws. In both cases, tremendous public good was served by the loss of some individual rights. He concluded with the striking and often-quoted sentence, "Three generations of imbeciles are enough."[17]

The effect of the ruling in *Buck v. Bell* and Holmes's strong language was significant to the ongoing attempts by proponents of compulsory sterilization laws. No longer were there any lingering doubts about the constitutionality of such legislation; arguments about their violations of due process, the prohibition on cruel and unusual punishment, or an unreasonable expansion of the states' police powers dissolved and were replaced with questions about the efficacy of compulsory sterilization. The annual number of sterilizations nationwide skyrocketed. In 1925, just under 6,000 compulsory sterilization had been recorded, but within ten years that number would top 20,000. On October 19, 1927, a few months after the publication of the Court's decision, officials at the State Colony sterilized Carrie Buck, her mother, and her daughter. Seventy-five years after the Bucks' sterilization, Virginia governor Mark Warner offered an official apology that denounced compulsory sterilization and his state's involvement with it as a "shameful effort."[18]

Catholics and Coerced Sterilization

The only Supreme Court justice to vote against the sterilization of Carrie Buck and her family was Justice Pierce Butler, who did not write a dissenting opinion to counter Holmes's strongly worded majority opinion. Lacking a direct statement from him, historians and legal analysts have been left to ponder the basis for his dissent, and some have pointed to the justice's religion. Butler was the only Catholic on the Court. Twenty-five years after the decision, an article in the *Catholic World* claimed that Holmes had told a fellow justice in reference to the case, "Butler knows this is a good law. I wonder if he will have the courage to vote with us in spite of his religion."[19]

The most recent analysis of possible motives for Butler's opposition to the majority opinion, Phillip Thompson's "Silent Protest: A Catholic Justice Dissents in *Buck v. Bell*," discussed the role of the justice's religious beliefs and quoted an earlier case in which Butler asserted, "religion cannot be separated from morality and . . . without it character will not be secure . . . against the attacks of selfishness and passions."[20] Biographers identified two other cases in which Butler's religion may have had some bearing on his decisions, *Pierce v. Society of Sisters*, which struck down the Oregon law that would have effectively closed Catholic schools and forced students into public classrooms, and *Cochran v. Louisiana Board of Education*, which permitted public money to be spent for the purchase of textbooks for private schools.[21] Both cases were decided unanimously, so "one might well conclude that Butler's religion was negligible in the Court's decisions."[22]

Despite the ease with which one might assume that Butler's dissenting vote in *Buck v. Bell* was motivated by his Catholic religious beliefs, it is just as likely that it emerged from his concern for individual freedom and for due process, which he had made clear in earlier cases and stemmed, one biographer argues, from his "keen sense of individual liberty" that emanated from his childhood roots in Minnesota.[23] Butler's commitment to individual rights and his efforts to limit the ability of courts to infringe on those liberties were evident in the volume published in his memory by the Bar and Officers of the Supreme Court of the United States after his death in 1939. The resolution at the front of the book emphasized Butler's belief that "there is a law greater than the judges," as well as his conviction to avoid misapplication of the law "merely because the end in view appeared at the moment to be desirable." In similar fashion, Chief Justice Charles Evans Hughes Jr. called Butler a "stickler for the rights of those accused of crime to be protected against the abuses of authority."[24] Clearly, among his colleagues Butler was well known to be powerfully motivated by an intense respect for individual rights and a sincere belief that the power of authorities needed to be held in check to avoid the destruction of individual liberties. It is just as easy to assume that his vote in *Buck v. Bell* was motivated by these convictions as by his religious beliefs.

Butler's concern for personal liberties in the *Buck v. Bell* case is further evidenced by the note that Chief Justice William Howard Taft sent to Holmes along with his request that Holmes write the majority opinion. Taft wrote, "Some of the brethren are troubled about the case, especially Butler." Taft suggested that to mitigate these concerns, which were not explicitly stated in the note, Holmes ought to emphasize the "care Virginia has taken in guarding against undue or hasty action, the proven absence of danger to the patient, and other circumstances tending to lessen the shock that many feel over the remedy." The recommended cures suggest that the source of concern rested in the justices' trepidation about violating Carrie Buck's civil liberties. Ultimately, Taft presaged Holmes's infamous line, "three generations of imbeciles are enough," by concluding, "The strength of the facts in three generations of course is the strongest argument."[25]

Perhaps the most convincing argument that Butler's vote in *Buck v. Bell* had little to do with Catholicism is founded on the fact that in the late 1920s, the church's position on sterilization was not perfectly clear. While there was a growing sense among American Catholic leaders that coerced sterilization in the name of social and biological progress was deeply problematic, there had been nearly two decades of statements on both sides of the debate from Catholic authorities. The lack of a clear Catholic position on eugenics in the first quarter of the twentieth century was especially evident in a series of articles in the *American Ecclesiastical Review* from 1910 to 1912.[26] Published in either English or Latin, the articles obviously were meant to be read by the clergy to help them consider the most appropriate message to present to the laity. Spurred by the initial article by Stephen Donovan, professor of moral theology at Catholic University in Washington,

D.C., theologians debated the question of the morality of sterilization surgeries undertaken "to counteract certain evils of inherited degeneracy."[27] Donovan was generally supportive of coerced sterilization, believing that it was "an easy and safe way of preventing hereditary physical and moral degeneration."[28] His position was quickly attacked by Monsignor De Becker of Louvain University and by P. Rigby, professor of theology at the Dominican College in Rome, who argued that "a surgical operation which involves a notable mutilation not necessary for the conservation of life is contrary to the moral law."[29] Within a few months, Donovan's position attracted supporters, among them an anonymous author who called himself Neo-Scholasticus and Theodore Labouré, a professor in the diocesan seminary of San Antonio; both directly addressed the claims of moral law and agreed with Donovan's call for Catholics to support the passage of eugenic sterilization laws.

By the summer of 1911, the theologians had come to a stalemate, and the editors of the *American Ecclesiastical Review* summarized the theologians' arguments and attempted to clarify the issues at hand. Both sides, they explained, held strong opinions as well as a number of assumptions not based on certainty. The editors noted that "there is disagreement among theologians, for example, regarding the effect of this operation as inducing the impediment of matrimony called *impotentia*."[30] Their summary and publishing of the moral analysis of the issue by the Innsbruck theologian Albert Schmitt instigated a second round of articles and letters on the subject, in which Austin O'Malley emerged as the chief opponent of the procedure. O'Malley's June 1911 "Vasectomy in Defectives" described the laws in effect in several states, tallied the financial cost of the insane to U.S. taxpayers (estimated to be $85 million annually), and described the procedure and effects in medical detail. O'Malley allowed for states to use sterilization as a form of punishment, but believed that the state had no right to mutilate citizens in the name of improving future generations. Doing so would turn governing into merely "human breeding and natural selection as applied to animals in a stockfarm raised for prize exhibition." He concluded by attacking "those who think they think scientifically" in arguing that "the State could wipe away all tears, cure disease and poverty by legislation" with "a snip of a pair of scissors in the hands of a gaol-surgeon, not omitting the fee for the snip."[31]

In England, the opinion among Catholics about eugenics was similarly unclear. Take, for example, the 1912 *The Church and Eugenics* by the Reverend Thomas Gerrard, which alternately praised and condemned aspects of the eugenics movement and concluded that it was "unwise either to approve or condemn the movement without many distinction[s] and reservations."[32] Firmly basing the impetus for the movement on Galton's and Pearson's work, Gerrard emphasized the eugenicists' human-animal analogy in arguing that "just as the animal can be improved by attention to heredity and environment so also can man be improved." He celebrated fellow Catholic Caleb Saleeby, who in his 1909 *Parenthood and Race Culture: An Outline of Eugenics* sought to "prevent the unfit from coming into existence," but argued that "once they are in existence we must make the best of them."[33] Ultimately, Gerrard believed that if the eugenics movement called for both hereditary and environmental improvements, it would be acceptable to Catholics.

The 1913 edition of *The Catholic Encyclopedia* likewise reflected the church's ambiguous position on the subject of eugenics and compulsory sterilization even more so than did the debate in the pages of the *American Ecclesiastical Review*.[34] The more than 1,600-word-long entry on eugenics devoted nearly half its space to a description of Galton's work and his general notions about heredity, spending very little time on the relationship of Mendelism to eugenics. The second half of the entry explored the similarities and differences between the church's position and that of the average American eugenicist. The author emphasized the fact that both eugenicists and the Catholic Church sought to eliminate "racial defects," and therefore the church "has no fault to find with race culture as such. Rather does she encourage it. But she wishes it carried out along right lines." The difference between the church's means and those of eugenicists centered on the fact that the church was most concerned with a person's "eternal life," whereas eugenicists were concerned with an individual's "civic worth." Moreover, there was some division between the two positions on the correct method to achieve their shared goal of preventing "defectives from propagating their kind." Eugenicists encouraged either segregation or sterilization, while the church preferred segregation because the operation "would open the door to immoral practices which would constitute a worse evil than the one avoided." Nonetheless, the entry emphasized the fact that eugenicists and the church were not far apart, as many eugenicists preferred segregation, and the Holy Office had not yet offered a clear opinion on the subject, so "the question is open." The entry's conclusion demonstrated the fact that for the first three decades of the twentieth century, even the church's most ardent opponents of coerced sterilization had to admit that "the Catholic Church has made no pronouncement on the question. In the absence of an authoritative decision from that source and in view of the division of opinion among the moral theologians, Catholics are morally free to adopt whichever view seems to them the more persuasive and reasonable."[35] There was in fact considerable exchange between Catholic authorities and American eugenicists between 1910 and 1930. Sharon Leon, who has analyzed American Catholic responses to the eugenics movement in great detail, explains that Catholics in the United States urged "American eugenicists to rid their movement of racial and class prejudice," and in so doing participated in a "revealing debate on immigration restriction, charity, racial hierarchies, feminism, birth control, and sterilization" that demonstrates numerous points of convergence between Catholics and eugenicists.[36]

Returning to the question of what motivated Butler's vote in *Buck v. Bell*, it must be recognized that before 1927, there was no clear Catholic position on eugenics or compulsory sterilization laws. While there were a number of ardent Catholic opponents to coercively sterilizing people deemed unfit to procreate, there was likewise support among some Catholic theologians for the laws.

Throughout the 1920s, the Roman Catholic Church was silent, and theologians were ambiguous on the issue, so Butler could not have been substantially influenced by the church's eventual position in opposition to compulsory sterilization.

The *Buck v. Bell* decision crystallized American Catholic opinion against compulsory sterilization laws. Take, for example, the editorial that appeared in the May 14, 1927, edition of *America*, less than two weeks after the case was decided. Calling the ruling "most unfortunate," the editors attacked "the tendency of Federal courts to set aside the deeper consideration of humanity and public policy in favor of conceptions that are purely legalistic." They did not dispute the legal arguments offered, and accepted the state's right to protect the common good. However, they maintained that there was no evidence whatsoever in support of the claim that the state of Virginia was in danger of being "swamped with incompetents," as Holmes contended in his majority opinion. Moreover, even if it were overrun with feebleminded citizens, they claimed that the state had no right to invoke "the extreme measure of sterilization" when "direful segregation affords all needed protection." Instead of improving the situation, sterilization invited serious evils and was another of the many "fallacious short-cuts to social health which have so often led us into the bog." Ultimately, their critique of sterilization was based on the Catholic belief that "every man, even a lunatic, is an image of God, not a mere animal, that he is a human being, and not a mere social factor."[37]

Throughout the latter half of 1927, the National Catholic Welfare Conference published three articles in its *Bulletin*, each motivated by the recent *Buck* decision and each harshly opposed to coerced sterilization. An editorial note explained that the articles were part of a "campaign to expose the fallacies of legalized human sterilization as the solution of the problems of the feebleminded and insane."[38] The first, "The Sterilization of the Feebleminded," written by A. R. Vonderahe, a specialist in the treatment of nervous and mental diseases, appeared in the August *Bulletin*. Vonderahe argued that sterilization was not "justified in the present status of our knowledge of feeblemindedness and the factors of heredity which it involves." Drawing from Abraham Myerson's and Walter Fernald's articles in the September 1924 edition of the *Archives of Neurology and Psychiatry*, he attacked eugenicists' claims that the feebleminded were especially "prolific in the matter of children."[39]

The following month, in September 1927, H. H. McClelland, former superintendent of the Dayton State Hospital for the Insane and director of the Ohio Association for the Welfare of the Mentally Sick, published "The Sterilization Fallacy" in the *Bulletin*. Admitting that eugenicists were probably "sincere in their contentions," he compared compulsory sterilization laws with prohibitionists' claims that "as soon as the saloon disappeared the asylums of the country would be emptied because alcohol was the great cause of mental sickness." Sterilization, he argued, "would be of no more value in eliminating the mentally sick than cutting off the ears of mental sufferers with the same objects in view."[40] Three months later, the *Bulletin* published an account of McClelland's talk before the National Council of Catholic Men Conference in which he claimed that "Darwin made it quite easy to jump to the conclusion that the real cause behind insanity was heredity," and that certain scientists "pounced upon the work of Mendel as being just the truth they were looking for."[41]

The *Buck v. Bell* decision also clearly motivated Charles Bruehl to publish in the spring of 1928 his *Birth-Control and Eugenics in the Light of Fundamental Ethical Principles*, which was stamped with the approval of the Censor Librorum and the archbishop of New York. Explaining that propaganda for "eugenics, birth-control and other unsavory schemes for racial improvement" would probably receive new impetus because of the Court's decision, he hoped his book would balance the "audacity of the propaganda and the speciousness of the arguments employed" by eugenicists. Bruehl began his book with an aggressive attack on birth control, countering claims made by advocates about how the increased availability and use of birth control would improve women's lives and strengthen families. Most of the work, however, was devoted to criticizing the relationship between eugenics and birth control and to arguing that "birth-control has so far worked dysgenically" because "desirable stocks" have not increased in number, and "no amount of exhortation can induce them to abandon their selfish practice." He also directly quoted Davenport's statements in *Heredity in Relation to Eugenics* that sterilization legislation "does not square with what we know about heredity."[42]

In sharp contrast to the debate in the pages of the *American Ecclesiastical Review* early in the second decade of the 1900s, by 1930 it was difficult, if not impossible, to find an American Catholic authority who continued to support compulsory sterilization, and whatever uncertainty there may have been regarding the Roman Catholic Church's opinion about eugenic sterilization before or shortly after the 1927 *Buck v. Bell* decision, the issue was entirely resolved within three years. The most effective American Catholic opposing coerced sterilization and helping to bring together opposition to sterilization was Monsignor John A. Ryan, a Catholic social theorist and advocate of progressive reforms. His pre–World War I work focused on applying Catholic standards of justice to the economic problems of the average American worker, and his first major work, *A Living Wage*, advocated minimum wage laws and labor unionization.[43] During the Great Depression, Ryan followed many other American progressives in advocating the reforms associated with Roosevelt's New Deal. When the immensely popular anti-Semitic Father Charles Coughlin turned against Roosevelt during the 1936 presidential campaign, Ryan came to the president's aid with an overtly political speech, "Roosevelt Safeguards America," which was broadcast nationwide.[44] Social reform, democracy, and the nature of civic and political responsibility dominated Ryan's life and work. He melded the works of Catholic authorities, like James Cardinal Gibbons and Archbishop John Ireland, with secular American, often Populist, authors, such as Richard Ely and William Lilly, to produce a

Catholic social gospel that demanded engagement with the nation's urban social and political problems.

Ryan was one of the few Catholic authorities who forged a successful relationship with American progressives. Even though Catholic Church leaders and members of the progressive movement shared a great many assumptions and goals, both sides found it difficult to overcome prejudices. The progressive movement, urban based and powerfully Protestant in both its worldview and constituency, was tinged with nativist beliefs about Catholic immigrants and a general mistrust of the Roman Catholic Church.[45] In the years before the turn of the twentieth century, Protestant clergymen like the Congregationalist Josiah Strong had typically criticized immigrants as immoral and criminalistic, assailing them for their adherence to Catholicism and socialism.[46] For their part, many American Catholics harbored resentment and distrust of the reforming progressives, bridling at the progressives' patronizing rhetoric and the civil service reforms that made it increasingly difficult for immigrants to use religious and ethnic ties to obtain jobs.[47]

Like Justice Butler, Ryan had been born and raised in Minnesota by Irish immigrant parents. Mentored by Archbishop John Ireland, a dominant figure in politics in and around St. Paul, Minnesota, Ryan left the Midwest to attend the newly created Catholic University of America in Washington, D.C., to study moral philosophy. He returned in 1902 to assume a teaching post at Saint Paul Seminary, then moved back to Catholic University in 1915, where he taught, headed the Social Action Department of the National Catholic Welfare Conference, and edited the *Catholic Charities Review*. "He was a man of seemingly boundless energy and endless causes. Even at the time of his death in 1945, he was still the dominant American Catholic social theorist."[48]

In 1930, shortly before Pope Pius XI made clear the Catholic Church's official position, Ryan entered the public discussion on eugenic sterilization with a twenty-page booklet titled *The Moral Aspects of Sterilization*, which summarized eugenicists' claims and synthesized for the first time the emerging Catholic argument against eugenic sterilization.[49] The work was part of a four-volume pamphlet series published by the National Catholic Welfare Conference that included descriptions of state sterilization laws, inheritance of mental defects, and social care of the mentally deficient.[50] Of the four, Ryan's *Moral Aspects of Sterilization* presented the most aggressive attack on eugenic sterilization from a theological point of view.

Ryan began his pamphlet by briefly summarizing the arguments for and against sterilization by the dozen Catholic authors who had participated in the discussion nearly twenty years earlier in the *American Ecclesiastical Review*. Ryan strongly emphasized that "all Catholic moralists" considered sterilization to be an acceptable practice when done to cure diseases, including treating "excess sexual erethism," an archaic term for someone who is abnormally excited by sexual activity. He attacked Labouré's claim that since the state could deprive "abnormal individuals of the exercise of other natural rights," the state was therefore justified in coercively sterilizing these same people. Ryan based his theological attack on sterilization on the "double effect rule," which prohibited an action that produced two effects—one good and one evil—when the evil effect is the cause of the good effect. "In the case of eugenic sterilization," Ryan explained, "the evil effect of the operation, namely, privation of the generative power, is the efficient cause of the good effect, namely the prevention of degenerate offspring. Therefore, it is morally unlawful." That is, while society as a whole gained if certain unfit citizens were sterilized, the means to the end were themselves evil, thus negating the potential benefit of preventing the perpetuation of a hereditary defect. Again and again, in analyzing the question of the moral acceptability of compulsory sterilization, Ryan concluded that the state was not within its rights in compelling individuals to give up their reproductive capacities, because to do so would violate a basic human right, the right to procreate. The other three volumes in the National Catholic Welfare Conference's pamphlet series also appeared in 1930, and each attacked compulsory sterilization laws as unconstitutional, immoral, and unjust. Intended for use in parish study clubs, all four pamphlets contained bibliographies and study guides. According to Leon, "study clubs helped to ensure that the adult lay population was well versed in the principles and reasoning of the church's theological, moral, and social teaching."[51] These groups would be invaluable in helping the Catholic laity confront the enthusiastic public claims about the potential of compulsory sterilization laws to solve complex social, political, and economic problems.

Despite the hard work of Ryan and the National Catholic Welfare Conference, until the Vatican offered a conclusive opinion on the subject at the end of 1930, the question of the moral acceptability of compulsory eugenic sterilization laws remained open for debate. On December 31, 1930, Pope Pius XI clarified the church's position on eugenic sterilization, as well as on a number of the issues related to marriage and sexual relations, with the landmark encyclical *Casti Connubii*, or *On Christian Marriage*.[52] Pius XI shed light on the official doctrine on birth control, which "had some ambiguities on the fine points of exactly under what conditions Catholics can try to avoid having children (including the legitimacy of the rhythm method), and some prominent church officials felt both European and American Catholics were quite ignorant of the doctrine."[53] *Casti Connubii* denounced the use of birth control to "deliberately frustrate" reproduction, and made clear that it was entirely appropriate for sterile husbands and wives to have intercourse because "the use of matrimonial rights" have beneficial secondary effects, "such as mutual aid, the cultivating of mutual love, and the quieting of concupiscence."[54] It also condemned abortion as against the precept of God and a violation of the law of nature, including abortions motivated by eugenic ideals. Finally, Pius XI ended the discussion over the church's position on eugenic sterilization by attacking "that pernicious practice." He claimed that there were some who were oversolicitous for the cause of eugenics and who put eugenics before aims of a higher order. Legislation that deprived citizens "of that natural faculty by medical action despite their unwillingness" was a state exercise in "power over a faculty which it never had and can never legitimately possess."[55]

As might be expected, Pius XI's 1930 encyclical was greeted warmly by American Catholics like John Ryan. The following year, Ryan published "The Moral Teaching of the Encyclical" in the *American Ecclesiastical Review*, condemning, as had Pius XI, state-ordered compulsive sterilization on the basis of the intrinsic sacredness of the human person.[56] In addition to encouraging Catholic authorities to speak out against eugenics in the 1930s, the encyclical spurred the generation of Catholic writers that followed it to attack the practice. Several years after the encyclical was issued and a month after Germany put into effect its sterilization law, the director of the National Catholic Welfare Conference Bureau of Publicity, Patrick J. Ward, evaluated the Catholic position on coerced sterilization in his article "The Grave Issue of Sterilization," which appeared in the national monthly magazine *Catholic Action*. Calling the German law "dictatorial" and "tyrannical," Ward distinguished the "feebleminded" from the "delinquent or criminally-minded," arguing that sterilization served no therapeutic purpose and caused a "great moral and public danger." Ward's greatest concern appeared to be the fact that authorities in the United States, Britain, and Germany had broadened their scope of targeted individuals to include not just the feebleminded and criminalistic but also "drunkards, drug addicts, epileptics, syphilitics, and those guilty of sexual crimes," as well as "schizomaniacs," the insane, blind, deaf and dumb, and those who suffered from St. Vitus's dance or a physical deformity. Especially in these cases, he argued, "the value of a human soul is of little consequence," and "science has lost its bearings."[57]

Ultimately, *Casti Connubii*, the National Catholic Welfare Conference's publications, and Ryan's efforts, along with the subsequent reception among American Catholics to all of these, crystallized the Catholic position against sterilization, both coerced and voluntary, and made the Roman Catholic Church the first nationwide institution to offer a widespread and organized resistance to compulsory sterilization in the United States. At first, the church's stance combined with a strong anti-Catholic sentiment to motivate some Protestant and secular leaders to continue their support of eugenic-based measures. It would take more than two decades for the position taken by Catholic clergymen, social theorists, lawyers, and scientists to begin to take hold in American culture. Once it finally did, the church's arguments aligned with both popular and professional American resentment of eugenics.

Throughout the 1940s and into the 1950s, Catholic theologians and scholars from several disciplines continued their attack on eugenics and compulsory sterilization laws. Edgar Schmiedeler's 1943 pamphlet, *Sterilization in the United States*, published by the same National Catholic Welfare Conference that had published Ryan's work a dozen years earlier, described the status of compulsory sterilization laws, summarized the relationship between heredity and eugenics, and concluded by examining the morality of sterilization. Myerson's work played a central role in Schmiedeler's analysis of the scientific consensus about heredity and eugenics, especially his attacks on the concept of feeblemindedness as unwisely broad in its conception. Schmiedeler also attacked the efficacy of compulsory sterilization by explaining that the notion that eugenic regulation could bring about a fit race was "sentimental delusion," because "even if every defective in existence were to be sterilized, this would not eliminate mental defect, as is sometimes thought; or even ever appreciably reduce its amount." His conclusions about the morality of sterilization were in line with those of Catholic theologians, and he stressed the role of environment and the potential of improvement in dealing with so-called feebleminded citizens. Nonetheless, positive eugenic initiatives, such as those undertaken by the Roman Catholic Church to increase the number of children born to fit parents, "have unquestionably done much to further race betterment."[58]

In 1944, Joseph Lehane, a Cleveland priest, completed a dissertation at Catholic University in which he summarized the morality of compulsory sterilization legislation, and he concluded that the laws were both scientifically invalid and morally reprehensible. Lehane emphasized the British foundations of modern eugenics, but made clear that the ideas on which the movement was based were ancient. His analysis accurately described the surgeries performed to sterilize both men and women, as well as other techniques, including the use of X-rays and hormone injections. Instead of beginning with American attempts to pass laws around the turn of the twentieth century, Lehane's legal history of compulsory sterilization described Mohammedan beliefs of the Middle Ages and the 1779 efforts by the German physician Johann Peter Frank to enact laws that would allow for the castration of the mentally diseased and mentally deficient in order to protect against the deterioration of humanity. In describing the scientific status of eugenics and the sterilization laws passed by its advocates, Lehane identified four basic claims made by sterilization proponents: feeblemindedness was increasing at alarming rates, defective individuals had children at higher rates than did normal people, heredity was the primary cause of both defects and social ills, and the environment was of little value in improving the heredity of the feebleminded. Lehane drew heavily from the American Neurological Association report of 1936 to attack most of the eugenicists' claims and concluded that heredity was probably of some, but little, importance as a causal factor in feeblemindedness and that the supposedly rapid increase in the number of defectives was grossly overstated.[59]

As did most other Catholic authors on the subject, Lehane balanced his attack on the scientific validity of eugenics with a discussion of the moral problems faced in the passage of compulsory sterilization laws. He concluded his book with a chapter on the relationship between sterilization laws and moral law, which emphasized a person's dual nature—endowed with an immortal soul as well as a material body. This duality required one to recognize that no matter how important heredity or environment was in the production of a person's material being, "any attempt to improve the human race that excludes the soul of man from consideration does not secure for itself an initial change of success because of the enormous gap in its premises." Lehane also challenged the power of the state to impose such drastic remedies on the bodies of individuals not guilty of any actual crime and called for a "more exact demarcation of the limits within which public authority may operate to secure its ends as a perfect society." Just as he respected the church's official position and attacked the scientific and moral validity of the negative eugenic policy of coerced sterilization, Lehane also followed his prior authors' habit of advocating positive eugenic initiatives in which the family is "fostered and safeguarded by society at large through a just political and economic regime." Both religious and secular officials ought to encourage the observance of "the laws of the married state, especially those relating to fecundity and fidelity," therefore "the eugenic program for Catholics must be almost entirely positive in nature."[60]

As other professions were just beginning to marshal opposition to eugenic sterilization, Catholic writers' animosity toward eugenics in general and compulsory sterilization in particular continued throughout the 1950s. More than twenty-five years after *Buck v. Bell*, J. E. Coogan, a Jesuit and director of the Department of Sociology at the University of Detroit, published "Eugenic Sterilization Holds Jubilee" in the *Catholic World*. After offering an ardent defense against sterilizing the three Buck women and summarizing Holmes's assertions in the majority opinion, Coogan attacked the practice of coerced sterilization, arguing that it "is enforced only among the poor; hence its exploitation of the defenseless usually escapes notice."[61] Marshaling quotes from J. B. S. Haldane, William McGovern, and Ashley Montagu, Coogan concluded with an argument that most assuredly would have held considerable sway in Cold War America: Holmes's words and the targeting of the poor for sterilization "sound too much like the sort of thing that Russian Communists delight to tell about us behind the Iron Curtain."[62]

In the 1950s, Catholic attacks on the power of the state to coercively sterilize some of its citizens intensified. Take, for example, the 1956 article by the Reverend Joseph D. Hassett, a Fordham University philosopher, titled "Freedom and Order before God: A Catholic View." Hassett discussed several biomedical issues from a Catholic point of view, including artificial insemination, the sale of contraceptives, and compulsory sterilization laws. Taking a hard line on all three subjects, he argued against the right of the state to coercively sterilize its citizens by claiming that people have "inalienable rights which cannot be forfeited to the state (even if the state usurps them by force) since the state has no legitimate claim on them." Addressing the counterargument that coercive sterilization is a tool for the state to protect itself against "the multiplication of defectives" that threatens its welfare, Hassett asserted that there is no such thing as the state, rather it is merely a union of individuals. Moreover, he added, "I would like to see convincing proof that the only way the state can protect itself against the criminally insane is by compulsory sterilization."[63] Compulsory sterilization laws were, to Hassett, immoral, illegal, and logically unsupportable. Nonetheless, sterilizations continued in many states, and the annual number of coercive sterilizations remained constant at over 1,000 per year throughout the 1950s.

A similarly uncompromising Catholic critique on eugenics and compulsory sterilization is found a decade later in the entry for eugenics in the 1967 *New Catholic Encyclopedia*. Written by Paulinus F. Forsthoefel, an expert in mouse genetics who would later author the book *Religious Faith Meets Modern Science*, the 1967 entry was of approximately the same length as its 1913 predecessor, but of considerably different content.[64] Forsthoefel briefly explained the origins of eugenics with Galton, moving quickly to deal with the American eugenics movement. Whereas the 1913 entry discussed eugenicists with more moderate views, such as Caleb Saleeby and Havelock Ellis, Forsthoefel focused instead on Davenport, Goddard, and Richard Dugdale. Likewise, there are significant differences in the scientific content of the articles, as Forsthoefel says nothing about Mendelism or the relationship between eugenics and genetics. Instead, he explores the role of eugenicists in bringing about compulsory sterilization laws and immigration restriction as well as in advocating birth control, especially "by those with inferior heredity." Reflecting the fact that 1930 was a watershed year in the history of Catholic opinion on the subject of eugenics and coerced sterilization, Forsthoefel wrote that "after 1930 eugenics in the U.S. and elsewhere rapidly declined," due in no small measure, he claimed, to the Nazis, who "carried racism to its logical conclusions." In sharp contrast to the 1913 entry's emphasis on the ways in which the church was likewise interested in race culture and in preventing "defectives from propagating their kind," Forsthoefel's entry concluded by condemning eugenicists' neglect for "man's supernatural destiny and its significance for his total life."[65]

The Most Dangerous Ally

In the 1930s, as the number of coerced sterilizations skyrocketed, the uncoordinated collection of people and groups who opposed eugenic sterilization began to coalesce. Their effectiveness was severely hampered by the difficulty they had in recognizing their shared goals. The greatest barrier to the emergence of a widespread antisterilization movement in the first half of the twentieth century was the inability of professional social scientists to come together with the only organized national resistance to eugenic sterilization, the Roman Catholic Church. The unwillingness of psychiatrists, sociologists, and biologists to work with the Catholics stemmed in large part from American anti-Catholicism and from many American scientists' distrust of the Roman Catholic Church. While the church was clearly instrumental in eroding support among a large segment of Americans for eugenic sterilization, one cannot overstate the tremendous impact of anti-Catholic sentiment in the United States and its role in undermining the Catholic Church's opposition to eugenic sterilization. In fact, it might well be argued that Catholic resistance to eugenics motivated many Protestant Americans to continue to support eugenics.[66]

In the years following the *Buck v. Bell* decision, dozens of scholars supported compulsory sterilization laws and wrote articles in both professional journals and popular magazines to shore up support for sterilization and to attack the Catholic Church's condemnation of it. Sociologists were generally supportive of compulsory sterilization laws, believing that biologists had adequately demonstrated the role of heredity in producing unfit citizens and accepting the state's responsibility to address the problem. For example, Frances Oswald's 1930 article in the *American Journal of Sociology*, which appeared to take a moderate stance on the need to sterilize the nation's mentally and physically inferior population, attacked the "obstacles in the path of eugenic sterilization." Emphasizing the "recent advances made by biologists with respect to the causes of mental and physical defects" and a realization of the significant role "played by heredity as a producing cause," Oswald cited Laughlin's *Eugenical Sterilization in the United States* to describe the sterilization operations in detail before discussing the various states that had passed compulsory sterilization laws. She identified the "intricacies of the law" and the "conservatism of American public opinion" as impediments to the operation of badly needed compulsory sterilization laws, outlining a number of court cases in which the laws were declared unconstitutional and describing Oregon's public referendum that invalidated the state's 1911 sterilization law. Every source of opposition, be it legal or popular, could be handled, except for one: "Statutes can be worded so as not to violate the Constitution, people can be educated to a broader outlook, but the Catholic church will remain firm in its opposition to sterilization." Focusing her attack on Charles Bruehl's 1928 *Birth Control and Eugenics*, Oswald summarized his arguments against sterilization and explained that the majority of Catholics "seem to agree with Bruehl that it is an 'unsavory scheme.'" Nonetheless, she concluded, "that there is a definite need for some sort of action to limit the number of degenerates has long been acknowledged by all thinking people."[67] Apparently, Oswald did not include Catholics among the "thinking people."

Perhaps the most aggressive attack on the Catholic opposition to compulsory sterilization came in Paul Blanshard's 1949 *American Freedom and Catholic Power*. Blanshard was an American journalist and Congregational minister whose best-known writings attacked the Roman Catholic Church as a dangerous and powerful institution that was undemocratic and therefore threatened the United States. He explained that he wrote his book because "American Catholics and American non-Catholics both tend to leave the discussion of religious differences to denominational bigots." Most important, he argued, was discussion of the "Catholic question," which was, he claimed, the fact that "the Catholic people of the United States are not citizens but *subjects* in their own religious commonwealth." The authority of the Roman Catholic Church exerted itself on American Catholics in such a way as to curtail the freedoms of Catholics and non-Catholics alike because of the influence that the church had on American culture. "There is no doubt," he wrote, "that the American Catholic hierarchy has entered the political arena, and that it is becoming more and more aggressive in extending the frontiers of Catholic authority into the fields of medicine, education and foreign policy." "As we shall see in this book," he explained, "the Catholic hierarchy in this country has great power as a pressure group, and no editor, politician, publisher, merchant or motion-picture producer can express defiance openly—or publicize documented facts—without risking his future."[68]

Blanshard's attack on the Catholic position on eugenics was framed in the context of Catholic opinions on sex and birth control. Drawing from papal encyclicals and popular Catholic periodicals, he argued that the Catholic Church's opposition to birth control was based on its philosophy of "conquest by fecundity." He claimed that the church's official stance was that "creating Catholics is a good thing in itself, and that even if they are diseased, feebleminded and a menace to normal community life, no *medical* act should be permitted to prevent their conception, their survival, or their freedom to produce other human beings." He attacked the "twisted and bizarre" principles of the Catholic sexual code, and asserted that it was the direct result of priestly celibacy, which produced the "restless pugnacity of the priests and the craving for authority."[69]

Catholic authors responded to Blanshard's claims by explaining that "the Roman Catholic position on birth control is based upon the natural law which applies to all men." Attacks on the notion of natural law, be they from Holmes or from Blanshard, rejected the notion that there was a set of laws governing morality and social interaction that was every bit as fixed and outside of cultural influences as were physical laws. For example, in his 1955 *Catholic-Protestant Conflicts in America*, John Kane explained that "there is no scientific evidence of a difference in Catholic and Protestant fecundity" and that Blanshard's claims were based on his unreasonable fear that "Catholics will reproduce so rapidly that they will become numerically the largest group in the United States." Defending the church's position on compulsory sterilization against Blanshard's attacks, Kane asserted that "the scientific case for sterilization is today a rather shaky one."[70]

Because of the significant anti-Catholic sentiment common in the United States in the first half of the twentieth century, the Roman Catholic Church's opposition to compulsory sterilization had limited effect. In states like Louisiana, which had a high proportion of Catholics, the church played a significant role in preventing the adoption of compulsory sterilization laws. In most places, however, it had little influence. For Catholic opposition to sterilization to effectively sway the American debate over compulsory sterilization laws, Catholic authors needed to secularize their arguments, something that would not happen until the mid-1950s. By the latter half of the century, the claims made by Catholic opponents to compulsory sterilization would be commonplace in the discourse over eugenics in the United States. The U.S. Supreme Court never overturned *Buck v. Bell*, but the 1942 U.S. Supreme Court decision in *Skinner v. Oklahoma* did substantially undermine the government's ability to coercively restrict individuals' reproductive abilities. Nonetheless, it did not have substanial impact on the practice of coerced sterilization in the United States, and coerced sterilizations continued well into the 1960s.

CHAPTER 5

The Professions Retreat

Beginning in the early 1930s, some of the American professions that supported eugenics and compulsory sterilization, including physicians, social scientists, and biologists, slowly withdrew their support. It took decades before widespread support for coerced sterilizations completely eroded and the word *eugenics* acquired its current negative connotations. After some early resistance from criminologists, who generally rejected hereditarian explanations for crime but accepted some aspects of the American eugenics movement and certain claims common among compulsory sterilization advocates, there are three identifiable sources for the eventual decline of support for compulsory sterilization in the United States and ultimately for the decline of American support for eugenics: a 1935 report by the Committee for the Investigation of Sterilization of the American Neurological Association, an article published in the *Georgetown Law Journal* that finally brought the longstanding Catholic opposition to compulsory sterilization laws into the mainstream, and a report by the Legal and Socio-Economic Division of the American Medical Association (AMA).[1] All three sources directly addressed the scientific justification for sterilization, the claim that certain undesirable traits were inherited and that compulsory sterilization could substantially reduce the number of people in the next generation with those traits.

How we today remember the American eugenics movement is every bit as significant to the declining support for coerced sterilization as was the abandonment of the movement by biologists, medical professionals, and social scientists. In this regard, the American historians and the historians of science who have written and continue to write this history have profoundly influenced our opinion of coerced sterilization. By the 1960s, just as the last professional support for eugenics was about to dissolve, a new generation of historians emerged and challenged longstanding notions about equality and the justifiable power of the state. They were increasingly critical of the progressives' willingness to restrict individuals' civil liberties as progressives sought to better society as a whole, and eugenical sterilization represented to late-twentieth-century historians one of the most egregious of governmental interventions. They alluded to parallels between the American eugenics movement and the atrocities committed by Nazi Germany and its allies during World War II, a narrative that came to its fullest development in Edwin Black's 2003 *War against the Weak: Eugenics and America's Campaign to Create a Master Race.*[2] It would certainly be deeply problematic to deny certain commonalities between American and German eugenics and ignore the fact that American progressives and the Nazis shared many of the same assumptions about the power that ought to be allotted to the state, the "social inadequacies" of certain citizens, and the potential benefits of carving out certain undesirable elements of society. However, imagining that American advocates of coerced sterilization were Nazi-like distorts both their activities and our ability to recognize contemporary authoritarian tendencies and the enduring influence of biological determinism in American culture. It also overlooks the fact that from the end of the World War II through the 1960s, there was no popular recognition of a link between the American eugenics movement and the Holocaust; this connection emerged in the 1970s. Claims that Nazi atrocities and a popular recognition of the link between eugenics and the Holocaust led to the immediate demise of eugenics and the immediate end of compulsory sterilization in the United States after 1945 are inaccurate.

The American eugenics movement appeared to be officially dead in the early 1980s—dead at least for the time being. The wave of high school and college textbooks published in the 1970s at first omitted any discussion of eugenics and coerced sterilization; by the end of the decade, they began critically evaluating and openly dismissing eugenics and compulsory sterilization laws as both scientifically and politically untenable. Many of these authors demonized American eugenicists by directly linking them to the Nazis, and they perpetuated the myth that postwar revelations of Nazi atrocities soured American public opinion against eugenics. Actually, when one considers the public uproar over coerced sterilization in states like Kansas, voters' rejection of compulsory sterilization laws via referendum in Oregon, and the dozens of trials to prevent the sterilizations of mental health inmates and prisoners, it could easily be argued that, outside of the 1930s, the majority of the American public never supported compulsory sterilization, but coerced sterilizations nonetheless occurred in most American states.

Early Professional Resistance to Compulsory Sterilization

In the years between 1910 and 1920, several articles appeared in the *Journal of the American Institute of Criminal Law and Criminology* that demonstrated criminologists' interest in the question of the use of compulsory sterilization laws to prevent, rather than merely punish, crime. In 1910, in the first volume of the journal, editors discussed plans at the forthcoming International Prison Congress to discuss methods of preventing crime. They claimed that efforts to check the increase in the number of criminals—including "care of discharged prisoners and their families, colonies for inebriates, prison schools and libraries, dealing with vagrancy and alcoholism, and kindred methods"—were necessary, but they came too late. They were "like fighting a prairie fire close up to the haystacks, with the wind driving the sparks ahead of the flame. We need 'backfiring' to keep the hungry enemy at a distance from home and harvest." Professionals needed to find a "humane and effective substitute for natural selection and capital punishment," such as "prolonged and progressive sentences for habitual, professional and dangerous criminals, more thorough segregation of the insane, epileptics and feeble-minded, colonies for inebriates, and a few advocate asexualization." The editors concluded, "All these methods deserve a place in the discussions" hosted by the journal, and over the next several years it published articles from both critics and opponents of compulsory sterilization laws.[3]

Throughout the second decade of the twentieth century, the *Journal of the American Institute of Criminal Law and Criminology* published articles both supporting and attacking the use of compulsory sterilization laws to address the problem of crime. Among the articles supportive of sterilization was a translation from Italian of Giulio Battaglini's "Eugenics and the Criminal Law," in which he stated that it was "natural to demand that measures be adopted to hinder the reproduction of those offenders who constitute deleterious racial elements" so as to prevent them from inflicting "upon society a posterity with criminal tendencies."[4] In 1913, the editor of the journal, Robert H. Gault, stated that while criminality itself might not be inherited, "there is, however, inheritance of predispositions which are responsible for delinquency and crime in succeeding generations."[5] Three years later, W. F. Gray wrote a letter to the editor of the journal arguing that it mattered little if "medical skill should fail to prove that the tendency to crime is inheritable; not one criminal out of ten can raise a large family (and their families are most always at large) without each child learning from, and imitating its father; and in all probability surpassing his father in cussedness. Whether he learns it or whether he inherits it from his parent matters nothing; the fact remains that he *has* it."[6]

In 1914, the American Institute of Criminal Law and Criminology created the Committee on the Sterilization of Criminals, which was devoted to the "ecclesiastical endeavor" of changing the emphasis of the social sciences from cure and reform to prevention. Consisting of judges, mental health professionals, and biologists, there were several prominent proponents of compulsory sterilization on the committee, including Bleecker Van Wagenen, H. H. Hart, Harry Sharp, William T. Belfield, and Harry Laughlin. The *Journal of the American Institute of Criminal Law and Criminology* published a comprehensive report from the committee on each state's sterilization laws and court challenges to them, and an official statement from the committee on the main issues involved requiring the sterilization of state wards who were feebleminded, insane, epileptic, or criminalistic. The committee reported that there was no clear agreement on the question of whether criminal traits were heritable. However, because criminality was based in feeblemindedness and other shortcomings, "it is granted by most authorities that the sterilization of individuals" who suffered from "feeblemindedness, epilepsy, insanity, imbecility, alcoholism, syphilis and other characteristics and diseases" would ultimately "reduce the amount of criminality in the next generation." Weighing the question of the best possible operation for those ordered sterilized, the committee offered support for both vasectomy and castration and concluded that "much more thorough research needs to be made before a satisfactory answer can be given to the question" of the advantage of one operation over the other. The report included the often-cited "Model Sterilization Law," which was drafted and publicized by the Eugenics Records Office. The model law was, the report concluded, "better than any of the statutes which have been passed as yet," and the committee advised legislators to consult it in the passage of future compulsory sterilization laws. Appended to the end of committee's report on sterilization was a short letter of protest from John Webster Melody, professor of moral theology at the Catholic University in Washington, D.C. Perhaps alerted to the report by one of the committee members, Father P. J. O'Callaghan from Chicago, Melody objected to the report because it ultimately supported compulsory sterilization laws. He argued that "it has not been proven that criminal tendencies are inheritable" and that segregation of the feebleminded appeared to him "to be the only practical plan for the protection of society." As discussed in the previous chapter, from the 1930s through the 1950s American Catholics offered the only significant organized opposition to compulsory sterilization laws, and Melody's arguments were among the first Catholic responses to the laws.[7]

Among the criminologists who published in *Journal of the American Institute of Criminal Law and Criminology*, even the advocates of compulsory sterilization saw it as a dramatic intervention that was fraught with legal, moral, and scientific problems. Take, for example, the 1914 article by Frederick Fenning, a lawyer from Washington, D.C.. In his analysis of recent court decisions regarding compulsory sterilization, he concluded that the courts would uphold sterilization "as proper means of placing legal restrictions upon procreation," and he hoped that ultimately "an actual positive betterment of the race" would occur in ways that "shall not do violence to the rights of any." It would not be the work of men like Galton, Sharp, and Laughlin, he argued, but rather that of "Ehrlich, Plaut, Alzheimer, Bonhoeffer, Biedel, Lewandowsky and their associates and disciples" that would "bring about the true enhancement of the public welfare." Fenning also expressed a sincere concern that such legislation could easily go too far, and asserted that the courts needed to "influence and control the zeal of the social welfare worker, as well as the activity of the surgeon."[8] In a similar fashion, in 1915 H. C. Stevens, director of the Psychopathic Laboratory at the University of Chicago, published "Eugenics and Feeblemindedness," in which he argued that it was too early to draw any valid conclusions about the Mendelian inheritance of feeblemindedness. "There is no justification whatsoever for considering feeblemindedness a unit character of the same sense that tallness or dwarfness of peas, or the color coat of guinea pigs, or brachydactilism in man may be considered unit characters [*sic*]." Instead, he asserted, feeblemindedness is a composite of psychophysical reactions, and there was no known cause of it. Far from rejecting eugenics because of these lacunae, though, he concluded by arguing that it had contributed "important results to the control of feeblemindedness."[9]

The first critical, professional evaluation of compulsory sterilization laws appeared in the journal in 1914. In "Inheritance as a Factor in Criminality," Edith R. Spaulding, the resident physician at the Reformatory for Women in South Farmingham, Massachusetts, and William Healy, the director of the Juvenile Psychopathic Institute in Chicago, reported their analysis of 1,000 cases of young repeat offenders in which they sought to establish inheritance as a factor in criminality. Their study considered two types of inheritance: "(a) the direct inheritance of criminalistic traits in otherwise normal individuals; (b) the indirect inheritance of criminalistic tendencies through such heritable factors as epilepsy, insanity, feeblemindedness, etc." The two authors unequivocally concluded that they found "no proof of the existence of hereditary criminalistic traits, as such." The idea, they argued, "of bare criminalistic traits, especially in their hereditary aspects," was "an unsubstantiated metaphysical hypothesis." They did, however, also conclude that crime was "indirectly related to heredity in ways most important for society to recognize"; namely, that indirect causes such as epilepsy, feeblemindedness, and psychoses were inherited and did in fact contribute to the development of criminalistic tendencies in individuals.[10] From shortly after 1910 onward, the report was cited by opponents of compulsory sterilizations laws; for example, in 1914 the *Michigan Law Review* published an editorial that stated: "All such legislation is based upon the theory that heredity plays a most important part in the transmission of crime, idiocy and imbecility. In spite of a great amount of statistics gathered upon the subject, there is no convincing evidence that criminality is transmissible. This is apparent from the investigation of Dr. Edith R. Spaulding . . . and Dr. William Healy."[11]

The harshest attack on compulsory sterilization laws that appeared in the *Journal of the American Institute of Criminal Law and Criminology*, and one of the first professional attacks on sterilization laws, was Charles Boston's 1913 "A Protest against Laws Authorizing the Sterilization of Criminals and Imbeciles." Boston exclaimed, "Someone's idea of the public weal is the excuse for every abuse ever committed by power!" He began the long article by comparing Indiana's 1907 sterilization law to the English Parliament's singling out of Richard Rouse to be boiled to death for poisoning seventeen of his family members. Boston attacked assumptions about heredity that were used to justify such laws and stated that "the suggestions which lead to the sterilization of criminals and imbeciles come from sociologists and amateur reformers, and not from biologists or students of heredity." The potential utility of the laws was questionable, he argued, and they represented little more than a "pseudo-reform" enacted because of the "demand of a dangerous, though sincere element in the community." Sterilization represented an unacceptable extension of the police powers of the state and constituted a cruel and unusual punishment for the commission of crime.[12]

In the years following Boston's criticism of compulsory sterilization laws, the journal published several other criticisms, including a summary of Lester Ward's 1913 attack on eugenics that appeared in the *American Journal of Sociology*.[13] Ward, a botanist, paleontologist, and sociologist, was the first president of the American Sociological Association and a harsh critic of laissez-faire advocates, especially those who employed Darwin's work to justify unrestrained competition. Ward contended that the group of theories that made up eugenics "are largely old popular fallacies in a new dress," and he criticized the assumption that "man knows better than nature how to guide the forces of heredity."[14] The journal also published F. Emory Lyon's 1915 attack on eugenic sterilization, "Race Betterment and the Crime Doctors." Lyon was the superintendent of the Central Howard Association in Chicago, a prison-reform organization, and was a staunch critic of surgical solutions to crimes. "Crime doctors galore have arisen in every generation and every country," he asserted, "to proclaim a sure specific for the eradication of antisocial conduct, and the prevention of delinquency." Comparing eugenic sterilization with E. H. Pratt's operation to sever the penile nerve or Serge Voronoff's grafting of a baboon's thyroid gland into a feebleminded boy, Lyon attacked both compulsory sterilization laws and their advocates.[15] In a similar fashion, in a 1926 critique of the legal, economic, and social status of epileptics that appeared in the journal, the physician L. Pierce Clark bluntly stated, "Sterilization constitutes cruel and unusual punishment" and was "an unwarrantable exercise of the police power of the states." He also reported that "it has been held on scientific grounds that sterilization is not based upon sufficiently well established data."[16]

A clear indication of the views of the members of the American Institute of Criminal Law and Criminology on compulsory sterilization came in 1917, when William White, the chairman of the Committee on the Sterilization of Criminals, called for the dissolution of his committee and was supported by a majority of the committee's members. White cited statements like Bleecker Van Wagenen's claim, "I do not believe in inherited criminality as a trait . . . and therefore see little use of studying sterilization as a remedy for crime independently of its association with true mental defect," and former chairman Joel Hunter's statement, "The more I find out about it the stronger my feelings become against the sterilization of criminals as such." White also quoted Harry Laughlin, who stated, "I think the committee should insist that it be excused from writing further opinions not based upon research" and went on to claim that "criminality as a unit trait is not inherited." White concluded that "it is quite evident that there is not unanimity of opinion among the members of the committee," and therefore "there is no further necessity for it being continued until scientific, statistical and social work has been completed by the various agencies not engaged therein." William T. Belfield, ever the proponent of compulsory sterilization, submitted a brief minority report requesting that the committee not be dissolved, but rather its members replaced "by men whose views on other topics do not incapacitate them for the study of a problem in public welfare." The committee was dissolved, and the nation's principal professional organization devoted to criminology officially, but quietly, withdrew support for the compulsory sterilization of criminals.[17]

Neurologists Confront Compulsory Sterilization

In the early 1930s, as the annual number of compulsory sterilizations was rapidly increasing, the American Neurological Association established the Committee for the Investigation of Sterilization. Its members—Abraham Myerson, James Ayer, Tracy Putnam, Clyde Keeler, and Leo Alexander—were charged with evaluating "in a critical manner both the facts and the theories which constitute the subject matter of the inheritance of mental diseases, feeblemindedness, epilepsy, and crime."[18] The committee's report, published in book form in 1936 as *Eugenical Sterilization*, completely ignored nineteenth-century physicians' advocacy of coerced sterilization and mistakenly asserted that the original justification for compulsory sterilization laws was Darwinian, "since the survival of the fit is made critical in the original Darwinian theory." This assumption demonstrated that Darwinism was employed in both supporting and in attacking compulsory sterilization, and critics asserted that Darwin was used to justify both progressive interventionism and laissez-faire competition. The "humanitarian trend" of the nineteenth century had been increasingly criticized, the report claimed, by those who argued "that conditions of modern civilization lower the birth rate of the better groups, increase the birth rate of the dregs of society and consequently, as social conditions become tinged with humanitarianism, they spell also the biological ruin of mankind."[19] However, the rise of "true genetics" required a reevaluation of the claims made by eugenicists, which the committee had undertaken. Compulsory sterilization laws were founded, the report explained, on propagandists' claims that there was a "substantial increase in the number of feebleminded, epileptic, paupers, alcoholics, and certain criminals." Examining statistics, the committee members concluded that "there is no real increase in the commitment rate" and that "the race is not rapidly going to the dogs."[20] Likewise, they dismissed claims made by advocates of compulsory sterilization laws that defectives reproduced at rates considerably higher than did fit citizens. When the issue of the increased death rate of the mentally ill and retarded was taken into account, they concluded, the supposed threat of rapidly breeding defectives was further undermined. The committee members concluded that, even if sterilization measures were effective, there was no threat that validated their application. Moreover, too much stress, they explained, "has been laid upon the expense of caring of the mentally ill" in institutions, which really was not that great and only a little more expensive than caring for them at home.[21]

The report concluded with a series of recommendations from the American Neurological Association. First, its authors explained, the study of human genetics had not yet matured enough to warrant something as drastic as the sterilization of those with "manic-depressive psychosis, dementia praecox, feeblemindedness, epilepsy, criminal conduct or any of the conditions which we have had under consideration." Second, there was not sufficient sound scientific evidence to justify sterilization "on account of immorality or character defect." Human beings and their actions, they asserted, were far too complicated and too interwoven with social conditions to be explained entirely on a genetical basis. "Until and unless heredity can be shown to have an overwhelming importance in the causation of dangerous anti-social behavior, sterilization merely on the basis of conduct must continue to be regarded as a 'cruel and unusual punishment.'" Finally, they concluded, a much greater study of the impact of the environment was necessary to justify claims about the root causes of personal and social disorders. "That scientific day is passed when the germplasm and the environment are to be considered as separate agencies or as opposing forces."[22]

Catholic Critiques Go Mainstream

After three decades of activism, public pronouncements, and edicts handed down from the Vatican through the clergy to church members, in the 1950s Catholic resistance to eugenic sterilization finally developed an appreciative audience among non-Catholics. The most significant turning point for the public's reception of the Catholic critique came in a twenty-four-page article in the *Georgetown Law Journal* by James B. O'Hara and T. Howland Sanks in 1956. The article was part of a series of studies on population trends and movements conducted by the Social Science Department of Loyola College of Baltimore that included two other articles on birth control that were published in *Eugenics Quarterly*. O'Hara's and Sanks's institutional affiliations with Georgetown and Loyola suggest that Catholic social thought played some role in the article's content and the nature of their arguments, which corresponded nicely to prior Catholic theologians' claims. O'Hara and Sanks offered a brief history of eugenic and punitive sterilization in the United States and a detailed survey of state laws allowing or requiring eugenic sterilization. They wrote that for "the past fifteen years, eugenic sterilization in the United States has been on a steady decline," although they noted there had been a sustained decline over only the previous four years, with a ten-year high of 1,526 sterilizations in 1950.[23] They concluded with a list of professional, social, and legal reasons why the use of compulsory sterilization was declining in the United States during the 1950s. They pointed to Abraham Myerson's American Neurological Association report of 1937, which recommended that existing laws be amended to allow for the sterilization of only certain well-defined groups. "This general 'go-slow' attitude of the American Neurological Association," O'Hara and Sanks claimed, "undoubtedly has had great influence in bringing about the gradual decline in the eugenic sterilization movement."[24] Its influence, they asserted, was augmented by a reexamination of beliefs among medical providers regarding heredity and a gradual rejection of the notion that social and physical ills ran in families.

Legal and religious thinking about coerced sterilization had also been changing over the previous decades. O'Hara and Sanks pointed to Pope Pius XI's encyclical letter *Casti Connubii* and its clear condemnation of coerced sterilization: "Public magistrates have no direct power over the bodies of their subject."[25] Protestant authors, on the other hand, appeared to offer "no significant opposition or support of the sterilization movement," according to O'Hara and Sanks. Nonetheless, "persistent opposition by some churchmen and moralists has helped to solidify public opinion against ready recourse to eugenic sterilization." Moreover, a "perceptible change" in legal thinking had accompanied the broader social changes.[26] Most notably, the 1942 *Skinner v. Oklahoma* case firmly established procreation as a fundamental right.

O'Hara and Sanks's article is notable for two contributions to the demise of the American compulsory sterilization movement. First, it translated Catholic protests about compulsory sterilization laws, which had originated a quarter century earlier, into secular arguments against them. Throughout the 1930s and 1940s, law journals rarely discussed sterilization laws, and when they did, the articles were generally supportive. However, most articles that appeared on the subject in American law journals in the 1960s and 1970s referenced O'Hara and Sanks's article and parroted their arguments. They effectively made the arguments against compulsory sterilization that were originally offered by Catholic theologians acceptable to non-Catholics. Second, it was one of the first academic publications that directly linked the demise of the American eugenics movement and compulsory sterilization laws with the emerging American recognition of the Holocaust. O'Hara and Sanks wrote: "The abuse of sterilization legislation in Nazi Germany was a tremendous factor in turning American public opinion against the whole concept of compulsory state action. This is apparent from the instances of those American eugenicists who, after being unrestrained in their earlier praise of the German laws, found themselves compelled to retreat to more conservative positions when the Nazis used eugenic sterilization as an instrument of genocide."[27] The 1960s marked a turning point in American conceptions about both eugenics and the Holocaust; just as Americans did not recognize the term *Holocaust* until well into the 1960s, the term *eugenics* acquired negative connotations around the same time, and the two were increasingly linked together in American memory.[28] In the 1970s, as biologists, historians, and legal experts wrote critically about the American eugenics movement, they followed the example set by O'Hara and Sanks and linked American eugenics directly to World War II atrocities.

American Physicians End Their Advocacy of Compulsory Sterilization

Drawing heavily from Albert Deutsch's 1949 *The Mentally Ill in America: A History of Their Care and Treatment from Colonial Times*, the AMA's Legal and Socio-Economic Division's 1960 report on compulsory sterilization in the United States marked the official end of American physicians' advocacy of sterilization laws. The report explained that sterilization laws had "moved with such rapidity that today many persons question whether this swift acceptance was wise from either a scientific or a legal point of view." The increase in the number of compulsory sterilization laws was caused, it asserted, by a confluence of three events in the late nineteenth and early twentieth centuries: "the launching of the eugenics movement by Sir Francis Galton, the re-discovery of Mendel's laws of heredity, and the development of simple, non-dangerous surgical techniques for the prevention of procreation." In their brief recounting of compulsory sterilization in America, the physicians ignored their predecessors' ardent support of it in the nineteenth century. According to the AMA's Legal and Socio-Economic Division, the proponents of eugenics and coerced sterilization were by and large those biologists who had adopted Mendelism: "It [Mendelism] was seized upon as being applicable to human beings. The proponents of this view decided that mental illness, mental deficiency, epilepsy, criminality, pauperism and various other defects were hereditary."[29] They also completely overlooked the very different notions of inheritance that their nineteenth-century predecessors had employed, which mingled biological and cultural heredity to justify sterilizing people who might pass along detrimental biological traits as well as those who would not raise their children in an appropriate environment.[30]

The AMA report concluded in much the same way as did the report from the American Neurological Association's Committee for the Investigation of Sterilization published a quarter century earlier. It called sterilization "a drastic remedy and generally a permanent infringement of bodily integrity." It asserted that citizens who had been coercively sterilized thus far had not been accorded reasonable protections from abuse of the procedure and that there were a number of illustrations of the disregard of basic civil rights in the application of the laws. The report also emphasized the claim that "scientific opinion differs as to the value of sterilization," and since court decisions had assumed that the conditions included in sterilization statutes are heredity in nature, "the constitutionality of such statutes is questionable if scientific opinion is divided concerning the effectiveness of this procedure." Finally, it concluded, Holmes's claim about "three generations of imbeciles" demonstrated his wit, but not his wisdom. With that, American physicians officially ended their century-long advocacy of compulsory sterilization laws, a movement that began with Gideon Lincecum in the 1850s.[31]

The Last American Eugenicists

Throughout the mid-twentieth century, as lawyers, physicians, historians, and social scientists were beginning to attack the concept of eugenics and the compulsory sterilization laws associated with it, American biologists continued to support it. Attacks on eugenics emerged in the 1930s and continued to slowly grow throughout the 1940s, 1950s, and 1960s. Beginning in 1933, a series of short articles that addressed the question, "Is Eugenics Dead?" appeared in the *Journal of Heredity*, published by the American Genetics Association, formerly known as the American Breeders Association. The editorial note that preceded the six articles explained that the association's council had for some time discussed the proper relationship between the association and the American eugenics movement. Members came to the conclusion that "the most useful purpose this organization could fill is to avoid the adoption of any definite and unequivocal 'policy' with regard to eugenics." This was based on the belief by most members that, given the present state of knowledge and the nature of public opinion, "there was great need for a source of accurate and unbiased information," rather than adherence to any particular program.

In response to the *Journal of Heredity*'s stated position on the advocacy of particular public policies, A. W. Forbes, a businessman from Worcester, Massachusetts, who had long promoted positive eugenic initiatives, wrote a short letter to the journal lamenting the fact that despite several studies that demonstrated the immediate need of a program of positive eugenics, nothing had been done over the previous several years. Forbes explained that because no practical policies had developed despite their obvious need, "the feeling is spreading today that eugenics is a fad that has come and passed, that it should be consigned to the realm of visionary utopias, unworthy of consideration by practical men."[32] In response, he suggested that the journal begin publishing such proposals so that they could be adequately judged, then put into action.

Forbes's call received responses from five American eugenicists, including the heads of the nation's three largest eugenics organizations, a director of a research institution, and a professor of zoology. All five agreed that Forbes had done the eugenics movement a great service by initiating a conversation over the appropriate public policy recommendations eugenicists ought to be offering. Most of the responses also emphasized the victories achieved in the passage of sterilization laws in most states. They all differed immensely, however, in their conclusions about precisely what should be done regarding the current state of the American eugenics movement. It is clear from their varied responses that by 1933, eugenics was clearly at a crossroads, having achieved considerable successes in some venues, while suffering roadblocks and in some cases setbacks in others.

Clarence G. Campbell, president of the Eugenics Research Association (ERA), offered the first and longest response to Forbes's letter. Campbell's organization presented itself as the research wing of the American eugenics movement, leaving propaganda in support of eugenic initiatives for the American Eugenics Society. He met Forbes's claims about the lack of an adequate eugenical program with a series of assertions about the need to recognize that modern social theory has no foundations in biology; it if did, policy makers would be forced to recognize that "evolution displays no tendency to strive for equality in individuals, but rather to produce variant individuals, and to eliminate those of inferior survival value." This fact, combined with a "coarse and unrefined stock-breeding attitude" that eugenicists were mistakenly believed to hold, had led the general public to conclude that any eugenic problem would "be an intrusion upon the personal liberties of individuals." Nonetheless, "the United States has the distinction," Campbell explained, "of being the first nation in modern times to enact laws for eugenic purpose." He pointed to the more than thirty states that had adopted eugenic sterilization laws, the "one or two states" that had laws against "miscegenetic marriage," and the role of eugenics in helping enact stricter immigration restrictions.[33] Of all the American advocates of eugenics and compulsory sterilization, Campbell is among the most unsavory. An overt racist, a staunch advocate of Hitler, and later closely associated with Holocaust deniers, he is one of the figures who provide an obvious link between American eugenicists and Nazi proponents of coerced sterilization. William Tucker, who has written on the history of scientific racism, described Campbell as the "Nazi Press's favorite non-German eugenicist" and quoted the *New York Times*'s description of him as a "champion of Nazi racial principles."[34]

Conversations like the one in the pages of the *Journal of Heredity* appeared occasionally from the 1930s through the 1960s, suggesting that American biologists continued, albeit sometimes self-consciously and later cautiously, to advocate certain aspects of the eugenics movement. How can we determine when or if American biologists finally abandoned their advocacy of eugenics in general and compulsory sterilization in particular? If individual biologists' claims about the potential application of genetics to improving the quality of the human gene pool are examined, one finds even today a handful of advocates of eugenics. There are no surveys of biologists throughout the twentieth century that polled their opinions about eugenics or about the coerced sterilization of people with genetic ailments.

In *The Structure of Scientific Revolutions*, Thomas Kuhn argued that textbooks are valuable demonstrations of a shift in scientific opinion, and the textbooks that American biology professors wrote and used in their undergraduate survey biology courses provide a valuable measure of American biologists' advocacy of eugenics and coerced sterilization.[35] Steven Selden's study of eugenics in American biology textbooks has made clear that college students were taught about eugenics and sterilization throughout the first half of the twentieth century.[36] How long did this last? An examination of 200 college-level biology survey textbooks reveals that support for eugenics and its associated remedies for human ailments and social problems emerged in the 1920s, grew steadily throughout the early 1930s, and remained strong until the late 1960s. As the number of textbooks published annually spiked dramatically in the early 1970s, an increasing number of texts ignored eugenics, followed shortly thereafter by the demise of support for eugenics and then open attacks on it.[37]

Looking closely at some of the specific claims offered in American biology textbooks, it is obvious that throughout much of the twentieth century, American biologists saw in eugenics both the potential for bettering humanity and a useful strategy for funding basic scientific research in genetics. For example, in his 1937 *Elements of Modern Biology*, Charles Robert Plunkett claimed that "on the basis of our present knowledge, . . . mental as well as visible characteristics are undoubted[ly] largely determined by genetic factors; but as to details—what and how many genes are involved in any particular case, and to what extent their effect may be modified by environmental factors, such as nutrition, disease, education, social and economic conditions, etc.—we still know very little."[38] In the 1940s, Michael Guyer published *Animal Biology*, in which he asserted that "such definite advances in our knowledge of the processes of human heredity are being made that we can no longer refuse to take up the social duties which the known facts thrust upon us."[39] In the early 1950s, in *Fundamentals of Biology*, Murville Jennings Harbaugh and Arthur Leonard Goodrich claimed, "Society would benefit greatly if well-planned programs of sterilization and segregation were practiced, and if marriage laws were sanely standardized and enforced."[40] A decade later, Garrett Hardin wrote that "mankind must now invent new corrective feedbacks to restore the equanimity of life" because people had "upset the primeval balance of nature by producing Pasteur and all that his name symbolizes."[41] By the early 1970s, when the number of textbooks published annually skyrocketed, authors began to openly criticize eugenics. For example, in *Contemporary Perspectives in Biology*, Robert Korn and Ellen Korn explained that the word *eugenics* has "a negative connotation because in the past, attempts were made to relate the science of human heredity to a social movement, despite the fact that few scientifically valid facts were available." This was a problem, they argued, because science had finally developed to a point where a "scientifically valid eugenics program" was feasible. "Numerous biologists," they concluded, "including some of the most noted and learned geneticists, feel that a comprehensive eugenics program is the only answer to the problem of the future genetic state of man."[42] Similarly, in *Concepts of Biology: A Cultural Perspective*, Neal D. Buffaloe and J. B. Throneberry explained that eugenics received "a bad name," and as "a result most geneticists turned away from human genetics as a whole. Now this trend is being reversed, and there is a tendency in genetics to review the human science in the light of new knowledge."[43]

Looking generally at the trend in American biology textbooks' discussion of eugenics, we see that biologists' advocacy of eugenics began in the 1920s. The percentage of textbooks that advocated eugenics continuously increased until the end of the 1960s, when the total number of textbooks published increased dramatically. Throughout the 1960s and 1970s, there was no critical discussion of eugenics or of compulsory sterilization in American biology textbooks; however, by 1970 the percentage of books that said nothing about eugenics increased rapidly. We do not begin to see criticisms of eugenics or of compulsory sterilization laws in biology textbooks until the late 1970s. Judging from an analysis of biology textbooks, it seems biologists finally abandoned the American eugenics movement around 1970, and within a decade they became harshly critical of it (see Figure 5.1).

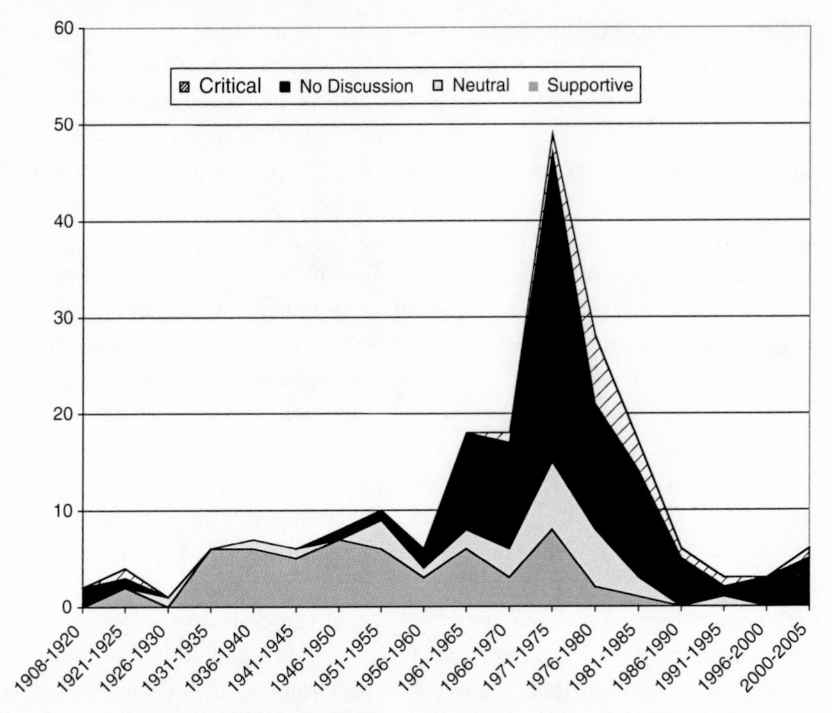

Figure 5.1. Eugenics in Biology Textbooks.

American biologists finally turned against eugenics and its associated remedies in the early 1970s, and they did so by making an explicit connection between the American eugenics movement and the Nazis. John Moore and Harold Slusher's *Biology: A Search for Order in Complexity*, which was supported by the Textbook Committee of the Creation Research Society, was the first American textbook to openly attack eugenics and to compare American eugenicists with the Nazis. An overtly creationist text, it stated, "Government control of eugenics has its dangers. Hitler exterminated six million Jews because he considered them 'unfit.' He also executed certain other people whom he regarded as inferior. Certainly, this cannot be condoned."[44] Linking evolution, eugenics, and the Holocaust is a common rhetorical tactic by opponents to the teaching of evolution in public schools and has been since the early 1940s.[45]

Although clearly not driven by the same creationist motivations that inspired Moore and Slusher, in the early 1970s a handful of biology textbook authors began attacking eugenics generally and made overt links between eugenicists and the Nazis. For example, in his 1971 textbook, Thomas Steyaert described how attitudes similar to those held by American eugenicists "were put into tragic practice by the Nazis, who killed millions and sterilized thousands who did not fit into their stereotype of Nordic supermen."[46] Three years later, in a section at the back of their *Fundamental Concepts of Biology* titled "The Authors' Views,"

Gideon Nelson and Gerald Robinson discussed the research of XYY males and concluded, "We should never forget Adolf Hitler's horrible and tragically erroneous attempt to create a 'pure' race of Germans by practicing genocide on Jewish people. It provides a frightening example of what can happen when political schemes are imposed on man's hereditary destiny."[47] Even those who still generally advocated eugenics in some form used the Nazis to make their point. In *Biology: A Full Spectrum*, Gairdner Moment and Helen Habermann contrasted the more humane negative eugenic initiatives like sterilization with the infanticide committed by the Spartans and the use of gas chambers by the Nazis.[48]

By the end of the twentieth century, American biology textbook authors extended their claims about the relationship between American eugenics and the Nazis to assert that the example of the Nazis' eugenics program turned Americans against eugenics. This is clearly seen in *Biology: Concepts and Connections* by Neil Campbell, Lawrence Mitchell, and Jane Reece, published in 2000. They cited James Watson's 1990 claims in support of the Human Genome Project that "we have only to look at how the Nazis used leading members of the German human genetics and psychiatry communities to justify their genocide programs." The authors went beyond Watson's assertion to state, "Largely because of the events in Nazi Germany, our society today rejects the notion of eugenics—the effort to control the genetic makeup of human populations. The possibility of gene therapy on germ cells raises the greatest fears in this regard."[49] Precisely how aspirations for genetic engineering of sperm and unfertilized eggs have anything to do with genocide is never explained within the textbook. Instead, it is merely offered as fact that most Americans fear therapies based on modern genetics because of the Holocaust. By the 1980s, perhaps because of their predecessors' hesitancy to reject eugenics and associated policies, the authors of American biology textbooks became some of the most assertively antieugenic voices in the United States.

Biologists' professional publications and organizations likewise provide evidence that the late 1960s represented the end of the biologists' support for the American eugenics movement. Between 1968 and 1973, the very same range of years during which authors of biology textbooks radically changed their position on eugenics and compulsory sterilization, the names of several prominent professional journals and an organization were changed to remove the word *eugenics* and to replace it with the words *social biology* or *biosocial science*. In 1968, the name of the *Eugenics Quarterly* was changed to *Social Biology*. A year later, *Eugenics Review* became the *Journal of Biosocial Science*. In 1973, the American Eugenics Society, the premier advocate of the study of human heredity and its social application, changed its name to the Society for the Study of Social Biology.

The end of biologists' advocacy of eugenics and coerced sterilization did not come easily, nor was it complete even by the 1970s. Even biologists who turned against the compulsory sterilization of those with genetic disorders still advocated unusual and problematic solutions. Take, for example, an article by Linus Pauling, two-time Nobel Prize winner and, as the Web site for the Linus Pauling

Institute proclaims, one of the world's "greatest scientists and humanitarians and a much respected and beloved defender of civil liberties and health issues."[50] In 1968, he wrote the foreword for the *UCLA Law Review*'s special issue "Reflections on the New Biology." The issue included articles on biomedical engineering, cryonics, organ transplantation, and experimental medicine. Pauling's introduction emphasized the momentous changes that had taken place in science over the previous half century and their impact on American society. Biology's most influential contribution, he argued, was the rise of molecular biology, which led to the discovery of the double-helical structure of DNA and medical advances. Pauling used the examples of sickle cell anemia and phenylketonuria to demonstrate how knowledge generated by molecular biologists required careful consideration by policy makers and promised amazing results. Identifying those citizens who were carriers of the gene that caused these diseases and preventing carriers from marrying would eliminate the ailments. Instead of advocating compulsory sterilization for these heterozygotes, Pauling offered a novel, albeit disturbing, solution: "I have suggested that there should be tattooed on the forehead of every young person a symbol showing possession of the sickle-cell gene or whatever other similar gene, such as the gene for phenylketonuria, that has been found to possess in single dose. If this were done, two young people carrying the same seriously defective gene in single dose would recognize this situation at first sight, and would refrain from falling in love with one another."[51]

An examination of the depiction of eugenics and of compulsory sterilization laws in American biology textbooks clearly demonstrates two things. First, if textbooks are to be considered an authoritative source of American biologists' beliefs on the subject, the profession did not stop advocating eugenics and compulsory sterilization until about 1970. Second, shortly after they abandoned the American eugenics movement, biologists began to aggressively attack compulsory sterilization laws as scientifically invalid. Their rapid turn against the movement, absent any fundamentally new scientific knowledge about heredity or evolution, suggests that social factors played an important role in their eventual opposition to coerced sterilization.

Writing the History of Coerced Sterilization

The first historian of coerced sterilization was Jacob Henry Landman, and he began the tradition for historians of offering aggressive criticism of coerced sterilization. From the late 1920s through the early 1930s, Landman, a professor at the College of the City of New York, wrote about coerced sterilization in articles that appeared in law review journals and in his 1932 book, *Human Sterilization: The History of the Sexual Sterilization Movement*.[52] Trained as a lawyer, Landman wove together blunt statements about the danger that criminals and feebleminded citizens presented to the nation with frank criticisms of the claims made by eugenicists, legislators, and judges. At a time when nearly every other American professional, including lawyers, medical doctors, criminologists, and most biologists, assertively advocated coerced sterilization, Landman was just as assertive in his opposition to it.

Despite buying into the more alarmist claims made by eugenic propagandists about the threat of feeblemindedness and the potential danger of hereditary criminality, Landman was nonetheless opposed to the use of sterilization as a solution to crime or mental deficiency. Often, according to Landman, coerced sterilization had the potential to make "defectives" even more dangerous because it allowed prison and asylum officials to release them from their institutions, which would allow them to "commit felonies . . . even murder." Women who were sterilized "would be more prone to engage in illicit intercourse, to adopt a life of prostitution, and spread venereal disease." He concluded, "In short, the interests of the community can only be adequately protected by the segregation of a considerable proportion of these persons in suitable institutions."[53]

Landman became interested in the subject of compulsory sterilization after reading Holmes's majority opinion in *Buck v. Bell*, believing that whatever legal foundations there might be in support of the laws, the "eugenic, social, and therapeutic values of this kind of legislation" were questionable. Nonetheless, later critics of coerced sterilization often find Landman's arguments problematic because, while he offered blunt criticisms of eugenic sterilization, he also accepted many of the alarmist claims about the threat posed by the mentally and physically unfit. For example, in a 1931 interview with the *New York Times*, he made exaggerated statements about the "socially inadequate" who "are a constant menace to our country and race." The paper quoted Landman as saying that "the chance of any one going insane, whether one be committed or not, . . . is at least one in ten."[54] Understanding the logic that underlies what at first appears contradictory in Landman's work requires later readers to recognize that, as did Davenport, Landman placed great stress on the need to conduct additional research on the subject. "The laws governing heredity are unknown to man," he argued. Much more work would have to be done by the emerging field of genetics before any laws or social policies could be based on it. He concluded, "To say the least, it is premature to base a monumental policy of sterilization on a belief that because a thing ought to be true, it, therefore, must be true."[55] This reasoning led him to call for a halt in the enforcement of eugenic sterilization laws "until knowledge of heredity was more reliable."[56]

Landman identified *Buck v. Bell* as a turning point in the history of coerced sterilization in the United States because, he believed, it would encourage more states to pass sterilization laws and end constitutional challenges to them in states where the laws already existed. He critiqued Holmes's opinion as "astoundingly brief and unusually platitudinous" and explained that he was "disconcerted by the absence of citations to support its legal principles." Nonetheless, he generally praised Holmes for this ability to "breathe an air of realism, humanity and progress" into his decisions as well as his willingness to grant "the several states greater jurisdiction as opposed to the federal government, so that the best interests of the people may be served." Even though he found fault with Holmes for "the lack of a thorough understanding of the field of eugenics," he ultimately laid blame on state legislators for *Buck v. Bell*. "The fault lies," he wrote, "with legislators who are incompetent to treat such profound and intricate fields of knowledge as heredity, eugenics and sterilization."[57]

From the late 1920s through the early 1930s, Landman focused on the subject of American compulsory sterilization laws, and he published two works on it: a significant journal article, "The History of Human Sterilization in the United States," which originally appeared in the *Illinois Law Review* and was republished in two subsequent journals, and a large book. The article was used in the 1929 Utah Supreme Court case *Davis v. Walton*, in which a Utah state prisoner, Esau Walton, was ordered asexualized.[58] On appeal to the Third District Court, Walton's attorney argued that the law violated his client's constitutional protections against cruel and unusual punishment and denied him equal protection of the law. Walton, an African American teenager, had been convicted of stealing several silk shirts and sent to the Utah State Prison. While there, a guard caught him having anal sex with another prisoner. Based on the guard's testimony as well as his claims that Walton had "frequently acted lovingly towards other boys who were confined in the prison," Walton was ordered asexualized. Precisely which operation was to be used—castration, severing of the nerve that controlled the penis, or vasectomy—was not specified by either the Utah law or the court. The outcome of the case most certainly would have made Landman happy: the court found that while the state was within its rights to asexualize Walton, "the record before us does not support the finding that by the law of heredity Esau Walton is the probable potential parent of socially inadequate offspring likewise afflicted." The court therefore struck down the case based on insufficient evidence to support the claimed outcome of the operation. Several years later, Landman proudly explained that the ideas he offered in his article were incorporated in the *Davis v. Walton* decision, which he described as "the prevailing law on the subject of human sterilization in this country today."[59] He was correct that his work was included, but he overstated the importance of *Davis v. Walton*, because *Buck v. Bell* would for years continue to be the most important decision on the subject of compulsory sterilization.

Landman's research and writing on the subject of coerced sterilization culminated in his 1932 book, *Human Sterilization*. Divided into five parts, the book addressed eugenics and American sterilization legislation, the attitude of the courts, the biological and eugenic bases for compulsory sterilization laws, the actual surgeries involved, and the difficulties legislators and judges confronted in attempting to formulate social policy based on the present knowledge of heredity. Much like Myerson's critiques offered about the same time, Landman emphasized the fact that scientists did not fully understand how human heredity worked, nor did they know precisely how, why, or even if feeblemindedness truly was hereditary. Even though the threats posed by the rapidly increasing number of criminals, mental defectives, and other feebleminded citizens were evident, scientists did not yet know enough to justify social legislation of this sort. Therefore, Landman concluded, compulsory sterilization laws were unwise at that time, no matter how badly they might be needed.

Reviews of Landman's book demonstrate the perils of taking a moderate position on a subject that crosses so many different disciplines. Kinder reviews, such as those that appeared in the *American Journal of Psychology* and the interdisciplinary social science journal *Social Forces*, praised the book as "a veritable storehouse of eugenic information" and for taking "a sane, conservative position in evaluating the eugenics movement in its historical perspective."[60] Reviews that appeared in legal journals were generally positive, emphasizing the book's comprehensiveness, and they tended to support Landman's conservative position on eugenic sterilization.[61] At worst, they accurately summarized his criticism of compulsory sterilization laws and countered that, despite any problems that might exist with them, if the laws prevented an "increase of mentally deficient persons," they were warranted.[62] In sharp contrast to such generally supportive reviews, however, was the one by sociologist and historian of contraception Norman Himes, of Colgate University, that appeared in the *Annals of the American Academy of Political and Social Sciences*. Himes called Landman's criticism "simply immature," and concluded that instead of demonstrating the futility of sterilization, "the author has shown only the futility of writing a self-contradictory, illogical treatise on sterilization."[63] It is difficult to determine if Himes was unhappy with Landman's criticisms because Himes supported coerced sterilization or if he was unhappy because Landman was not critical enough of it. An analysis of Himes's 1936 *Medical History of Contraception* helps clarify the question: he endorsed the views of several different advocates of eugenic sterilization and stated that even though "much of the earlier talk on heredity was nonsense," it should be recognized that "genetic knowledge has advanced considerably in the past two decades and fifty years from now we may know something about the subject. But this should not prevent us from acting in the interim according to our best lights."[64] Himes's support for eugenic sterilization is further evidenced by his choice of authors for the book's foreword, the gynecologist Robert Dickinson, who several years earlier had written an article for the *Journal of the American Medical Association* that supported eugenic sterilization.[65] Clearly, Himes's attacks on Landman's moderate position and criticisms of coerced sterilization were motivated by his support for eugenic sterilization as part of his overall advocacy for birth control.

A similarly critical review of Landman's book was offered by Edward Byron Reuter, a sociologist at the University of Iowa and president of the American Sociological Association, in the *American Journal of Sociology*. He harshly attacked Landman's moderate position, labeling Landman a conservative eugenicist because he was among those "who would sterilize only those defective persons who have inherited their defects and may transmit them (biologically) to their offspring," but not sterilize the "socially incapable." For Reuter, reproduction by those who were patently unfit to be parents was deeply problematic, and he assailed Landman's notion that such targeted sterilization would do anything to reduce the number of defectives. He called "childish" Landman's advocacy of the "sterilization of hereditary defects while opposing the sterilization of the congenitally defective who are equally incapable of competent parenthood."[66] It is not clear from Reuter's essay whether he opposed Landman's so-called conservative position and thus disliked his advocacy of even the most limited amount of coerced sterilization, or whether his true interest was preventing the unfit from becoming parents and thus found questionable Landman's unwillingness to sterilize those who would make incompetent parents. Analyses of some of his other published writings suggest that Reuter believed that eugenics was of no scientific value and that "heredity is a less important factor in populating quality than was once believed."[67]

By far, the harshest evaluation of Landman's *Human Sterilization* was offered by Samuel J. Holmes, a professor of zoology at the University of California and long-time advocate of eugenics, in a review for the *Journal of Criminal Law and Criminology*. After complimenting Landman's chapters on the present legal status of sterilization and the impressive compilation included in the book, Holmes disparaged Landman's "rather naive and even erroneous statements" on subjects such as Morgan's work on chromosomes, sex-linked characters, and the presence-absence theory, as well as his use of vague terms. "To dwell upon Dr. Landman's errors and misconceptions as to the genetics of mental defects and diseases," Holmes wrote, "would require more space than, I fear, would be allowed for this review." Nonetheless, Holmes devoted half of the long review to attacking Landman's "confused treatment" and his "lack of grasp of the modern factorial conception of heredity." The result, he concluded, "is that, while some of his criticisms are justified, he wastes much time in tilting against windmills."[68]

Throughout the first half of the 1930s, Landman himself was often tapped to review books on sterilization, and he frequently complied. For example, in 1935 he reviewed C. P. Blacker's *Voluntary Sterilization* for the *Journal of Criminal Law and Criminology* and attacked it as disingenuous for its recommendation of voluntary sterilization for mental defectives.[69] He gave a similarly negative review that same year of Leon Whitney's *The Case for Sterilization*, calling it a weak defense for voluntary sterilization. "The book," he explained, "is neither a scholarly nor scientific study, but a popularization of what meagre knowledge we have on the subject." Nearly half of the review consisted of a series of questions about how Whitney proposed to "educate the idiots, the imbeciles, the morons, and the criminals to subject *voluntarily* to sterilization," and how eugenicists could defend their calls to sterilize the unemployed "when society itself, in so many instances, is to blame for their economic insecurity."[70] His 1938 review of J.B.S. Haldane's *Heredity and Politics*, on the other hand, was strongly supportive of the author's contentions. Landman praised Haldane's "proposition that our current knowledge of human heredity does not justify the bio-social legislation which is supposed to be based on it," as well as his criticisms of "compulsory human sterilization laws of the United States and the anti-Jewish legislation of Germany." Haldane's book, Landman asserted, should be read by "all alarmist eugenicists."[71] By the late 1930s, Landman moved away from the subject of eugenic sterilization and grew increasingly interested in early-twentieth-century world history.[72]

After Landman's work in the early 1930s, it took three decades before American historians again turned their attention to the history of eugenics and coerced sterilization in the United States. In the 1960s, two works appeared: Mark Haller's 1963 *Eugenics: Hereditarian Attitudes in American Thought* and Donald Pickens's 1968 *Eugenics and the Progressives*. Both explored the nature/nurture debate and stressed the influence of hereditarian thought on the American eugenics movement. Of the two, Haller's book received far better reviews and was much more influential. In the 1970s, Kenneth Ludmerer's *Genetics and American Society* examined efforts by eugenicists to influence immigration restriction legislation during the 1920s. Also in the 1970s, Garland Allen began publishing articles on the subject with an essay on the relationship between genetics and class and with an institutional history of the Eugenics Record Office. His work, both as an author and as a supportive senior colleague, over the last thirty years has been instrumental in establishing a cadre of scholars interested in the history of eugenics. In the 1980s, Daniel Kevles published *In the Name of Eugenics: Genetics and the Uses of Human Heredity*, which set the tone for studying the history of eugenics throughout the 1980s and 1990s. Kevles examined eugenics in the United States and Britain, and he focused on issues like the importance of race and class as well as compulsory sterilization and immigration restriction legislation. As the book's title suggests, Kevles maintained that much had been justified "in the name of eugenics," including coerced sterilization. He also brought the story into the latter half of the twentieth century and argued that human geneticists had taken over the research done earlier by biologists as part of the American eugenics movement.

In 1991, Philip Reilly published *The Surgical Solution: A History of Involuntary Sterilization in the United States*, the first book since Landman's *Human Sterilization* devoted entirely to the history of compulsory sterilization. It was a valuable contribution to understanding coerced sterilization, and it brought the subject to the attention of many scholars who had previously overlooked compulsory sterilization in the United States. Reilly, a physician, geneticist, and attorney, was the executive director of the Shriver Center for Mental Retardation in Waltham, Massachusetts, and a specialist in legal issues raised by genetics. As had Kevles, Reilly assertively linked sterilization to only eugenics, and he set the tone for later analyses of coerced sterilization in the United States. Galton, Weismann, and of course Davenport and Laughlin were identified as the primary advocates of coerced sterilization in the United States. Reilly devoted only four paragraphs to the American physicians who advocated sterilization in the nineteenth century, and he began his story with Sharp's work after the turn of the century.[73] Unlike the hundreds, perhaps thousands, of American physicians that Reilly ignored, no American biologist ever sterilized a single mental health patient, prisoner, or welfare recipient. By overlooking the role of the American medical community and stressing only a handful of biologists, Reilly cemented the link between the long-dead eugenics movement and coerced sterilization.

By the turn of the twenty-first century, there were literally dozens of books devoted to the history of compulsory sterilization and the American eugenics movement. Nearly every one of them is critical of the scientific mistakes eugenicists made and the gross violation of civil liberties they advocated in their quest to improve the nation's gene pool. Some have taken a sensationalistic approach by purporting to uncover "the secret history of forced sterilization" or "America's campaign to create a master race."[74] Increasingly, historians have sought to lay bare the connection between earlier eugenicists and present popular notions about heredity and social deviance, and have effectively demonstrated how many of the assumptions on which the movement was founded are still with us today.[75] The most notable outlier in this discussion is Richard Lynn's 2002 *Eugenics: A Reassessment*, which laments the "rejection of a theory that is essentially correct."[76]

CONCLUSION

The New Coerced Sterilization Movement

Beginning in the mid-nineteenth century and continuing today, a number of deeply problematic assumptions about certain citizens' supposed social inadequacies have allowed for the coerced sterilization of tens of thousands of mental health patients and prisoners. In many cases, state wards signed permission forms, but the coercive nature of institutional settings is obvious, and it is difficult to defend the operations as truly voluntary. While physicians had campaigned throughout the latter half of the nineteenth century for the legal authority to sterilize defectives at will, it was not until the beginning of the twentieth century that legislators seriously considered passing such laws. Between 1907 and 1937, legislators adopted the claims made by an impressive collection of social and natural scientists, health care providers, mental health administrators, and prison authorities, and eventually two-thirds of American states passed compulsory sterilization laws. Based on available records, at least 63,000 state wards were coercively sterilized. The actual number is most certainly higher—perhaps much higher—as some physicians sterilized without state oversight, but because of limited documentation, the total number cannot realistically be estimated.

To anyone considering the subject around 1980, it would have seemed that the movement to coercively sterilize certain citizens had ended sometime during the previous decade. The withdrawal of support by many professionals—including physicians, legislators, biologists, and social scientists—combined with the increasing recognition of the racist and sexist motivations for the movement made involuntary sterilization too problematic to be advocated as a solution for any ailment, physical or social. For example, throughout the 1960s and 1970s, feminists publicized the coerced sterilization of women in the United States as well as in Puerto Rico, India, Bangladesh, and Brazil.[1] Poor women, they explained, were particularly vulnerable, especially in efforts to reduce the number of people on welfare.[2] Moreover, claims by American Indian activists that the

Indian Health Service had sterilized "at least 25 percent of Native American women who were between the ages of fifteen and forty-four during the 1970's" certainly should have effectively ended legislators' and physicians' ability to coercively sterilize anyone.[3]

Attacks on coerced sterilization also became part of the civil rights and black nationalist movements during the 1960s and 1970s. Take, for example, *Genocide in Mississippi*, published in the mid-1960s by the Student Nonviolent Coordinating Committee (SNCC). The twelve-page pamphlet was a reaction against a bill introduced in the Mississippi House of Representatives that would have "penalized the birth of an illegitimate child by imposing a prison sentence of 1 to 3 years on the parents" and was amended to allow "sterilization in lieu of the prison sentence." The representatives who sponsored the legislation, according to the pamphlet's authors, "made no attempt to disguise the anti-Negro nature of the bill." The pamphlet included not only a reproduction of the bill in question, but also the names and home addresses of every representative who voted for it.[4] More recently, Dorothy Robert's *Killing the Black Body: Race, Reproduction, and the Meaning of Liberty* explored the "meaning of reproductive liberty to take into account its relationship to racial oppression."[5]

While it is evident that coerced sterilization continued well through the 1950s and that there is little evidence to support claims that the Holocaust turned Americans against compulsory sterilization laws after the war, the comparisons between the American eugenics movement and the policies of the Nazis finally wielded powerful rhetorical force in the last decades of the twentieth century. Throughout the 1970s and 1980s, biologists and historians alike frequently linked coerced sterilization to the World War II atrocities. The 1960s and 1970s witnessed a major revision in many aspects of American history, including radically different views of the history of eugenics and, as demonstrated in Peter Novick's *The Holocaust in American Life*, of Nazi atrocities.[6] Two decades of Cold War anxieties and the growing conflict in Vietnam generated increasingly critical analyses of America's past, including the role of progressives in advocating aggressive interventionist policies, especially eugenics. Richard Hofstadter's criticism of social Darwinism in the 1940s and 1950s grew into a widespread critique of governmental attempts to cultivate a better American population. Mark Haller's *Eugenics: Hereditarian Attitudes in American Thought*, "the first comprehensive history of the rise, fall, and gradual revival of the eugenics movement in the United States," made an explicit link between American eugenic policy and the Nazis, who "demonstrated the uses that might be made of some of the eugenics doctrines" when they "stripped the eugenics movement of its trappings of science and disclosed that it had been based upon often careless and inaccurate research, that it was permeated by a virulent nativism without basis in fact, and that it frequently mirrored the conservative and reactionary social philosophies of its adherents."[7] Later authors of both U.S. history and biology textbooks routinely accepted and promoted claims about the

Nazi-like nature of compulsory sterilization laws specifically and the American eugenics movement generally.

Beginning in the late 1960s, biology textbooks often claimed that compulsory sterilization laws were ill-advised on scientific grounds based on the Hardy-Weinberg principle, which states that "in a large population in which random mating occurs, with respect to a particular pair of alleles, the frequency of the genes or alleles remains the same providing that there is not mutation, selection, or differential mating."[8] In considering attempts to eliminate or substantially reduce the frequency of an allele in a population, the principle asserts that as recessive traits grow increasingly rare, the ability of natural selection to eliminate them slows accordingly. Take, for example, Helena Curtis's 1975 *Biology*, in which she explained that the Hardy-Weinberg principle "is of special interest to students of eugenics," and used the example of diseases caused by a homozygous recessive gene, such as phenylketonuria. It would take about 100 generations, or roughly 2,500 years, to reduce the incidence of the disease from 1 in every 10,000 children born to 1 in every 40,000 children born. "The lack of effectiveness of such a program," she concluded, "is obvious."[9] Despite these claims, as Diane Paul and Spencer Hamish have persuasively demonstrated, very few of the geneticists who advocated compulsory sterilization failed to understand the implications of the Hardy-Weinberg principle.[10] Nonetheless, their support for the movement remained intact and often motivated them to call for even more aggressive measures to improve the overall genetic qualities of their nation's citizenry.

The decline in coerced sterilizations was not, in fact, caused by scientific advances like the Hardy-Weinberg principle, nor was it the direct result of World War II atrocities. Rather, it was brought about by the impact of political developments that took place during 1960s and 1970s. Among the most influential of these were the civil rights movements, specifically movements focusing on race, gender, sexual orientation, class, and physical and mental disabilities, as well as the efforts by advocates to protect the rights of prisoners and mental health patients.[11] These were the original sources, the fountainheads, from which widespread opposition to coerced sterilization emerged. Scientific, legal, and historical justifications—which ranged from analogies to the Nazis to the Hardy-Weinberg principle—followed, as did significant court decisions in the 1960s and 1970s. In contrast to Holmes's opinion in *Buck v. Bell*, later decisions affirmed citizens' right to control their reproductive abilities; the most notable of these are *Griswold v. Connecticut* in 1965, which overturned Connecticut's laws forbidding the use of artificial contraceptives, and *Roe v. Wade* in 1973, which legalized abortion in the United States prior to the point at which the fetus is viable.

Coerced Sterilization Today

Collectively, the court cases, various social pressures, scientific claims, and increasing political activism should have put to bed the notion that federal or state governments had a compelling interest in the reproductive lives of their citizens. For a time, they nearly did, and the movement to use sexual surgeries to solve complicated social problems appeared to have died near the end of the 1960s. However, occasional compulsory sterilization laws continued to be debated throughout the closing decades of the twentieth century, and some new laws were passed. Beginning in the mid-1980s, the issue of coerced sterilization reemerged as a number of state courts and legislatures revisited the notion of sterilizing certain citizens; namely, those men who had been convicted of particular sex crimes, women who abused their children, and "welfare queens," or single mothers on public assistance. Clearly, underlying assumptions about "social inadequacy," as Laughlin as had called it decades earlier, are still with us.

In 1969, state legislators in Montana passed a voluntary sterilization law that contained a clause aimed at those "who if they should procreate offspring might be expected either to transmit mental deficiencies to such offspring or be unable to adequately care for or rear such offspring without the likelihood of adverse effects on such offspring caused by such environment."[12] The law allowed the state eugenics board to decide whether mentally defective candidates for sterilization were capable of giving informed consent. All of those sterilized under the law were patients in the Boulder River School and Hospital, who were ordered sterilized before they were released into the community. Dr. Philip Pallister, clinical director of the institution, explained, "As a matter of institutional policy, we advocate voluntary sterilization as the only feasible means of birth control," because "self-control and planning [are] well beyond the capacity of most retardates." Pallister also claimed that when two retarded persons produced a child, the risk of bearing a retarded child was one in five. Officials at other institutions criticized the state's policy on the grounds that it was necessarily coercive. Dr. Judith Rettig, assistant commissioner of the division of mental retardation at the Department of Mental Hygiene for New York, asserted, "There is a very clear policy [within the profession] that sterilization should not be done because most retardation is not genetic." Likewise, Dr. James Clements, director of the Georgia Retardation Center who helped defeat a similar bill in his state, explained that "there is always the question of intimidation. Of course, you can be intimidated on the outside, too, but it's not the same as when you're under 24-hour surveillance." Between 1969 and 1974, at least sixty-four people were sterilized in Montana under the 1969 law.[13]

Ten years later, in 1979, Montana's sterilization law was revised to allow for sterilization in cases in which a patient's ability to properly consent was hampered by diminished IQ. It focused on those cases in which the patient might produce offspring who "might be expected to either: (a) transmit mental deficiencies to such offspring; or (b) be unable to adequately care for or rear such offspring without the likelihood of adverse effects on such offspring caused by such environment." The state legislature repealed its compulsory sterilization law in 1981.[14] By then, at least 320 people were sterilized under Montana's eugenic laws.[15]

In the last decades of the twentieth century, a number of men were targeted for punitive coerced sterilizations, especially as punishment for child molestation and rape. Throughout the 1980s and 1990s, several state legislatures considered bills that would require chemical castration or would have allowed surgical castration of convicted sex offenders. Four states enacted such laws—California, Texas, Florida, and Montana. The development of Norplant and Depo-Provera, the brand names of two female synthetic hormones, brought about a resurgence in the movement to control the sexual activities and reproductive capacities of some citizens. When used by women, Norplant provided long-term birth control that, unlike the birth control pill or mechanical birth control technologies, required nothing from the user after a medical professional had inserted a device containing the drug under her skin. Depo-Provera likewise required application by a medical provider, in this case the administration of a shot, and it prevented conception for three months. In men, these drugs caused what came to be known as chemical castration by making them impotent. Newspaper reports of the discussions surrounding adoption of the laws emphasized the inability of these men to resist their urges to rape and molest as well as the cost savings of castrating them as compared to imprisoning them.[16]

In 1990, Washington led the nation in considering the use of sexual surgery to reduce crime. The state legislature considered two bills, one that would have required castration of all sex offenders and another that would have allowed sex offenders to "choose castration in exchange for reducing their sentences by as much as 75 percent." Motivated by recent crimes, including the sexual mutilation of a seven-year-old boy, sponsors of the bill claimed that castration could cut repeat crime by 75 percent or more. After passing in the Senate and receiving broad public support, both bills were ultimately rejected by the House.[17]

In 1996, the California legislature adopted the nation's first chemical castration law, which required the chemical castration of repeat sex offenders. Governor Pete Wilson signed the law, saying, "Hopefully, this treatment will help in the difficult struggle to control the deviant behavior of those who stalk our young."[18] Legislators in Florida, Georgia, Iowa, Louisiana, Montana, Oregon, Texas, and Wisconsin soon followed with similar bills.[19] The 1997 Montana law that allowed for the chemical castration of men convicted of rape or incest stated that judges could impose on repeat offenders injections of Depo-Provera to reduce testosterone levels and sex drives. State representative Deb Kottel sponsored the bill and claimed, "It's like a nicotine patch. It takes the edge off and allows people to quit." Legislators were motivated to pass the bill by the apparent cost savings, as injections cost less than half of what it cost to incarcerate offenders.[20]

Throughout the 1970s and 1980s, judges, sometimes in partnership with the office of the district attorney, occasionally offered convicted criminals lesser sentences in exchange for their "consent" to be sterilized. Given the obviously coercive nature of such agreements, it is difficult to characterize these operations as truly voluntary. For example, in 1975 California Superior Court Judge

Douglas P. Woodworth approved the castration of two men, both forty-five years old, who had been convicted of child molestation and sentenced to life in prison. The men had requested castration to avoid lifelong prison sentences. State medical officials spoke out against the surgeries, and no doctor was willing to perform the operations. Stating that he had no other choice given their inability to find a willing surgeon, Judge Woodworth sentenced them to prison "with personal regret," having wanted to see the men castrated and put on probation.[21]

In 1984, South Carolina Circuit Court Judge C. Victor Pyle Jr. gave three men convicted of raping and torturing a woman the choice of thirty years in prison or castration.[22] At their sentencing, all three men seriously considered the option of surgery, but only one, Roscoe Brown, accepted it. In a *New York Times* editorial critical of Judge Pyle's offer of castration in lieu of prison, the editors mistakenly claimed that castration had never been used as government-sanctioned punishment. Their error was understandable given the widespread ignorance of the history of eugenics and compulsory sterilization in America. On appeal to the South Carolina Supreme Court, Pyle's offer was judged unconstitutional. The ineffectiveness of punitive and therapeutic castration was demonstrated in South Carolina when a man who had been chemically castrated because he had been convicted of rape in Texas was linked to seventy-five new sex crimes in the Richmond, South Carolina, area.[23]

In 1986, Debra A. Williams of Columbia, South Carolina, "submitted to sterilization to get a lesser charge in a plea agreement." Williams, a twenty-six-year-old woman who had been charged with murder for starving to death her twelve-week-old son, pled guilty to voluntary manslaughter. The killing had not been an isolated incident. Nine years earlier, Williams had been convicted of involuntary manslaughter in the death of her newborn daughter, and in 1981 she was convicted of assault and battery for beating her four-month-old son. The *New York Times* reported, "It was her own idea, according to her lawyer, Douglas S. Strickler, deputy Richland County public defender." The prosecutor consented to the plea agreement following Williams's sterilization because she "represented a threat only to her own children and her operation removed that threat."[24]

At least one Georgia district attorney and superior court judge believed that sterilization was a useful tool in punishing or treating criminals. In 2004, Carisa Ashe, the mother of eight, was charged with killing her youngest child while suffering from postpartum depression. Fulton County district attorney Paul Howard proposed a plea bargain that would allow Ashe to avoid a murder trial and a possible prison sentence if she agreed to be surgically sterilized. Superior Court Judge Rowland Barnes, after questioning Ashe to ensure that her agreement to be sterilized was voluntary, ordered her to serve five years' probation and undergo a tubal ligation within three months.[25]

In the 1990s, some Texas legal authorities considered castration as a treatment for sex criminals. In 1992, Steve Allen Butler, twenty-eight, was charged with sexually assaulting a thirteen-year-old girl. Butler had been on probation for molesting a seven-year-old girl, and he agreed to be castrated and serve ten years' probation rather than face a trial that could result in a life sentence. Butler's wife reportedly agreed to the procedure, but "doctors involved in the treatment of sex offenders and advocates of victims' rights . . . criticized the castration of sex offenders as a simplistic and questionable solution to a complicated problem." The court approved the deal, but Butler changed his mind before the operation took place. His relatives informed the press that he had been "manipulated, humiliated, intimidated, coerced, and brainwashed to the idea of castration by his attorney, Clyde Williams, and Judge Michael T. McSpadden." Race was an issue in the case. Butler was black, and leaders from the African American community, including Jesse Jackson, attacked the court, claiming that the offer for Butler to trade his testicles for his freedom smacked of the "social demongrelization schemes devised by eccentric right-wing lunatics and intentionally aimed at African-Americans." In the end, Butler stood trial for child molestation and received a life sentence. McSpadden, who had levied unorthodox sentences in the past, said, "Until we can live in something other than a constant state of fear, it seems that it would be altogether appropriate to attempt to render sexual offenders less capable of repeating their crimes."[26]

In 1996, the Texas legislature took up the subject of sexual surgeries for repeat child molesters. The case that motivated their interest involved a self-described "scum of the earth," Larry Don McQuay, who claimed to have molested young children at least 240 times. He first began serving a prison sentence in 1990 after being convicted of molesting a six-year-old boy. Unlike the earlier Butler case, McQuay was white, and he pushed for his own castration absent any discussion of a reduced sentence. Shortly before his release in 1995, he wrote to a victims-rights group that he was "doomed to eventually rape then murder my poor little victims to keep them from telling on me." McQuay requested "treatment," consisting of castration, to prevent him from molesting children in the future. He even attempted to castrate himself with a razor.[27] Then-governor George W. Bush said that he "supported castration and that he thought it could be accomplished without spending any taxpayer funds. There are going to be ample volunteers willing to contribute money to see that it's paid for."[28] McQuay was released without being castrated, but returned to jail in 1997 to serve a twenty-year sentence for molesting a boy several years earlier. That same year, the Texas legislature enacted a law allowing convicted sex offenders to volunteer for castration on the following conditions: the subject was a repeat offender, was at least twenty-one years old, provided informed consent, received proper psychological evaluations, agreed to participate in a ten-year follow-up study, and was appointed an independent monitor to ensure understanding of the process.[29] When McQuay was released in 2005, his lawyer reported that prison officials had ordered and carried out his castration. As of 2006, the law allowing repeat child molesters to volunteer for castration was still valid, and interested prisoners could apply to the Texas Department of State Health Services for the procedure.

The History and Future of Coerced Sterilization in the United States

While some legislators, social reformers, and judges continue to advocate sexual surgeries to solve complex and vexing social problems, members of the professions that had so ardently campaigned in favor of coerced sterilization in the nineteenth and twentieth centuries generally refuse to support it today. College biology textbooks, when they discuss the subject, firmly denounce the American eugenics movement and the associated compulsory sterilization laws. When judges in Texas, Colorado, and California recently allowed convicted sex criminals to reduce their prison sentences by submitting to sterilization, they could find no doctor willing to perform the operation. When legislators considered requiring mothers on welfare or felons to take synthetic female hormones, legal scholars attacked the ideas in their journals, nearly universally rejecting the claim that the state had a right or a compelling interest to require the use of the drugs.[30] Nonetheless, the advent of these new technologies means that many of the professionals who were once vital to the practice of coerced sterilization in the United States were themselves of less importance to the debates, because a surgeon is no longer necessary for the elimination of a person's reproductive capacity. The increasing capacity of biotechnologies and the advance of scientific knowledge make even more pressing the question of how they are employed to better the nation or to oppress citizens.

Perhaps the most startling and widely discussed example of the resurgence of the sterilization movement in the United States is Children Requiring a Caring Kommunity (CRACK), which has since been renamed Project Prevention for Children Requiring a Caring Community. The group represents the emergence of private sector interest in coerced sterilization and demonstrates that the coercion to "consent" to a sterilization procedure can originate from nongovernmental sources.[31] The organization was founded in 1997 by a housewife after she adopted four children from the same drug-addicted mother. It offers $200 to any drug-addicted woman willing to be sterilized or to receive long-term birth control like Norplant or Depo-Provera. After initial success in California, it opened satellite offices in Chicago and New York and attracted a considerable number of donors, including a well-publicized $5,000 gift from conservative talk show host Laura Schlessinger. Given the nature of drug addiction, a $200 payment is highly coercive. Hundreds of women took the offer, motivated both by Project Prevention's promotional slogan, "Don't Let a Pregnancy Interfere with Your Drug Habit," and by a need to support their addiction.[32] The organization currently offers cash incentives to women who are addicted to drugs or alcohol if they are willing to be sterilized or use any one of a number of long-term birth control options. If they receive a tubal ligation or Norplant, Project Prevention will pay the client $200. For Depo-Provera or an intrauterine device (IUD), participants receive $50 every three months.[33]

The social pressures and prejudices that brought about the movement to sterilize the nation's sexual and moral perverts, criminals, epileptics, and feebleminded are clearly still with us. Assumptions about the hereditary nature of what Harry Laughlin termed "social inadequacies," the incredibly powerful influence of professionals in reifying the public's bigotry, and the police power allowed to the state have in no way dissipated over the last century. In fact, in many cases, these forces have grown even stronger.[34] Improvements in scientists' understanding of heredity and the emergence of powerful new biotechnologies have also failed to undermine the sources of social influence that helped bring about the sterilization of tens of thousands of Americans. If we should learn anything from America's history of coerced sterilization, it is the absolute necessity of safeguarding our civil liberties. Much of the current literature on the subject of neo-eugenics, also called neugenics, attempts to elucidate how population control, genetic engineering, prenatal testing, selective abortion, and even genetic counseling are in fact operating on the very same assumptions that fueled the movement to enact compulsory sterilization laws.[35]

As the history of the American eugenics movement and of coerced sterilization has been written over the last four decades, there has been a strong tendency to demonize eugenicists and advocates of compulsory sterilization in such a way as to make them seem so alien, so out of the ordinary, and at times so nonsensical that it is easy to regard them and their crusade as an aberration. Overt connections between American eugenicists and the Nazis have only widened this gulf. By the end of the twentieth century, what little most Americans knew about the eugenics movement, they generally associated with fascist Germany and with the Holocaust. Despite occasional news stories and a growing body of literature on the subject, few Americans actually know about the nation's experience with eugenics and coerced sterilization; most simply cannot imagine that it could have ever happened in the United States. The demonization of notable eugenicists only exacerbates the problem. Recent histories, like Nancy Gallagher's *Breeding Better Vermonters* or Alexandra Minna Stern's *Eugenic Nation*, seek to make sense of eugenicists' motivations, and they demonstrate the continuity between us and the earlier advocates of coerced sterilization. This approach is vital to appreciating the fact that we still hold many of the assumptions that produced such horrible results. As Gallagher argues, "Our best safeguard against the injustices of the past rests with our willingness to confront our connection to this history prior to disowning it and with our recognition of the enduring power of research findings and the consensus of experts over our perceptions of other people's problems."[36] We must rescue the American eugenics movement and the advocates of compulsory sterilization laws from the dustbin of history—not to celebrate their prejudices or apologize for their mistakes, but to confront our connection with them. We need to appreciate our relationship with the eugenicists if we want to begin to challenge some of our own deeply problematic assumptions about other people's supposed social inadequacies.

The history of coerced sterilization in the United States reveals a series of unanswered questions, and we will not answer them until we appreciate the problematic assumptions on which compulsory sterilization laws were based. What precisely is wrong with coercively sterilizing some members of society? What claims have been useful in motivating physicians to coercively sterilize patients and in encouraging legislators to allow or even require their sterilization? There is also, of course, the ever-present question of how we balance personal liberties with the demands of public safety or the desire for apparent social or biological improvement. Perhaps the most troubling questions, especially to policy makers, are how and when do we take action in the face of uncertainty and risk. How complete and unchallenged does the scientific evidence have to be before we can justifiably infringe on personal liberties? Our ignorance of how prior generations of professionals approached these questions seriously hinders our ability to judge the best approaches. Worse yet, it prevents us from realizing that many of the prior assumptions that motivated the advocates of compulsory sterilization are, in fact, still with us today.

Appendix. Bibliography of Twentieth-Century American Biology Textbooks

Liberty Hyde Bailey and W. M. Coleman. *Animal Biology; Human Biology, Parts II and III of First Course in Biology*. New York: Macmillan, 1908.

James Needham. *General Biology: A Book of Outlines and Practical Studies for the General Student*. Ithaca, N.Y.: Comstock, 1912.

Truman Moon. *Biology for Beginners*. New York: Henry Holt, 1921.

Leonas Burlingame, Harold Heath, Ernest Martin, and George Pierce. *General Biology*. New York: Henry Holt, 1922.

Edward Menge. *General and Professional Biology with Special Reference to Man*. Milwaukee: Bruce, 1922.

Lorand Woodruff. *Foundations of Biology*. New York: Macmillan, 1922.

John Geisen and Thomas Malumphy. *Backgrounds of Biology*. Milwaukee: Bruce, 1929.

William Eikenberry and R. A. Waldron. *Educational Biology*. Boston: Ginn, 1930.

John Johnson. *Educational Biology: The Contributions of Biology to Education*. New York: Macmillan, 1930.

George Scott. *The Science of Biology: An Introductory Study*. New York: Thomas Y. Crowell, 1930.

Otis Caldwell, Charles Skinner, and J. Winfield Tietz. *Biological Foundations of Education*. Boston, New York: Ginn, 1931.

Arthur Baker, Lewis Mills, and William Connor. *Dynamic Biology*. New York: Rand, McNally, 1933.

Henry Barrows. *General Biology: A Textbook for College Students*. New York: Farrar and Rinehart, 1935.

Henry Barrows. *Elements of General Biology*. New York: Farrar and Rinehart, 1936.

Charles Plunkett. *Elements of Modern Biology*. New York: Henry Holt, 1937.

George Hunter, Herbert Walter, and George Hunt III. *Biology: The Story of Living Things*. New York: American Book, 1937.

Perry Strausbaugh and Bernal Weimer. *General Biology: A Textbook for College Students*. New York: J. Wiley and Sons, 1938.

Clarence Young, G. Ledyard Stebbins, and Frank Brooks. *Introduction to Biological Science*. New York: Harper, 1938.

Clarence Young, G. Ledyard Stebbins, and Clarence Hylander. *The Human Organism and the World of Life: A Survey in Biological Science*. New York: Harper and Frothers, 1938.
Amram Scheinfeld and Morton Schweitzer. *You and Heredity*. New York: Frederick A. Stokes, 1939.
Howard M. Parshley. *Biology*. New York: J. Wiley and Sons, 1940.
Michael Guyer. *Animal Biology*. New York: Harper and Row, 1941.
Frank Jean, Ezra Harrah, and Fred Herman. *Man and His Biological World*. Boston, New York: Ginn, 1944.
Douglas Marsland. *Principles of Modern Biology*. New York: Henry Holt, 1945.
Leslie Kenoyer and Henry Goddard. *General Biology*. New York: Harper, 1945.
Ernest Stanford. *Man and the Living World*. New York: Macmillan, 1945.
Truman Moon, Paul Mann, and James Howard Otto. *Biology: A Revision of Biology for Beginners*. New York: Henry Holt, 1946.
Laurence Snyder. *The Principles of Heredity*. Boston: D. C. Heath, 1946.
E. Grace White. *A Textbook of General Biology*. St. Louis: C. V. Mosby, 1946.
James Mavor. *General Biology*. New York: Macmillan, 1947.
Lorand Woodruff. *Foundations of Biology*. New York: Macmillan, 1947.
Lorand Woodruff. *Animal Biology*. New York: Macmillan, 1948.
Andrew Stauffer. *Introductory Biology*. New York: D. Van Nostrand, 1949.
Curt Stern. *Principles of Human Genetics*. San Francisco: W. H. Freeman, 1949.
Gairdner Moment. *General Biology for Colleges*. New York: Appleton-Century-Crofts, 1950.
Claude Villee. *Biology: The Human Approach*. Philadelphia: Saunders, 1950.
P.D.F. Murray. *Biology: An Introduction to Medical and Other Studies*. New York: Macmillan, 1950.
Garrett Hardin. *Biology: Its Human Implications*. San Francisco: W. H. Freeman, 1952.
Garrett Hardin. *Biology: Its Principles and Implications*. San Francisco: W. H. Freeman, 1952.
Lorus Milne and Margery Milne. *The Biotic World and Man*. Englewood Cliffs, N.J.: Prentice-Hall, 1952.
Murville Harbaugh and Arthur Goodrich. *Fundamentals of Biology*. New York: Blakiston, 1953.
William Hovanitz. *Textbook of Genetics*. Houston: Elsevier Press, 1953.
Paul Weisz. *The Science of Biology*. New York: McGraw-Hill, 1954.
Douglas Marsland. *Principles of Modern Biology*. New York: Henry Holt, 1955.
Thomas Hall and Florence Moog. *Life Science: A College Textbook of General Biology*. New York: Wiley, 1956.
Douglas Marsland. *Principles of Modern Biology*. New York: Henry Holt, 1957.
George Gaylord Simpson, Colin Pittendrigh, and Lewis Tiffany. *Life: An Introduction to Biology*. New York: Harcourt, Brace, 1957.
Claude Villee. *Biology*. Philadelphia: Saunders, 1957.
Lorus Milne and Margery Milne. *The Biotic World and Man*. Englewood Cliffs, N.J.: Prentice-Hall, 1958.
William Whaley. *Principles of Biology*. New York: Harper and Row, 1958.
P.D.F. Murray. *Biology: An Introduction to Medical and Other Studies*. New York: St. Martin's Press, 1960.
Relis B. Brown. *Biology*. Boston: Heath, 1961.
Garrett Hardin. *Biology: Its Principles and Implications*. San Francisco: W. H. Freeman, 1961.
Herbert Stahnke. *Biotic Principles*. Columbus, Ohio: C. E. Merrill Books, 1961.

Claude Villee. *Biology*. Philadelphia: Saunders, 1962.
William Bernard Crow. *A Synopsis of Biology*. Bristol: J. Wright, 1964.
Lawrence Dillion. *The Principles of Life Science*. New York: Macmillan, 1964.
Karl Von Frisch. *Biology*. New York: Harper and Row, 1964.
William Whaley. *Principles of Biology*. New York: Harper and Row, 1964.
A. M. Winchester. *Biology and Its Relation to Mankind*. New York: Van Nostrand Reinhold, 1964.
Douglas Marsland. *Principles of Modern Biology*. New York: Holt, Rinehart and Winston, 1964.
Alvin Nason. *Textbook of Modern Biology*. New York: Wiley, 1965.
Alfred Elliott and Charles Ray. *Biology*. New York: Apple-Century-Crofts, 1965.
Lorus Milne and Margery Milne. *The Biotic World and Man*. Englewood Cliffs, N.J.: Prentice-Hall, 1965.
Douglas Penny and Regina Waern. *Biology: An Introduction to Aspects of Modern Biological Science*. Toronto: Pitman, 1965.
George Gaylord Simpson and William Beck. *Life: An Introduction to Biology*. New York: Harcourt, Brace and World, 1965.
William Telfer and Donald Kennedy. *The Biology of Organisms*. New York: J. Wiley, 1965.
A. M. Winchester. *Modern Biological Principles*. Princeton, N.J.: Van Nostrand, 1965.
Cleveland P. Hickman. *Integrated Principles of Zoology*. St. Louis: Mosby, 1966.
E. Lendell Cockrum, William McCauley, and Newell Younggren. *Biology*. Philadelphia: Saunders, 1966.
Willis Johnson, R. W. Laubengayer, and L. E. DeLanney. *Biology*. New York: Holt, Rinehart and Winston, 1966.
Mary Gardiner and Sarah Fleminster. *The Principles of General Biology*. New York: Macmillan, 1967.
William Keeton. *Biological Science*. New York: Norton, 1967.
Robert Platt and George Reid. *Bioscience*. New York: Reinhold, 1967.
Neil Buffaloe and J. B. Throneberry. *Principles of Biology*. Englewood Cliffs, N.J.: Prentice-Hall, 1967.
Paul Weisz. *The Science of Biology*. New York: McGraw-Hill, 1967.
Hans Joachim Bogen. *Modern Biology*. London: Weidenfeld and Nicolson, 1968.
Jeffrey Baker and Garland Allen. *A Course in Biology*. Reading, Mass: Addison-Wesley, 1968.
John Alexander Moore. *Biological Science: An Inquiry into Life*. New York: Harcourt, Brace and World, 1968.
Helena Curtis. *Biology*. New York: Worth, 1968.
William Taylor and Richard Weber. *General Biology*. Princeton N.J.: Van Nostrand, 1968.
John Kimball. *Biology*. Reading, Mass: Addison-Wesley, 1968.
Charles Heimler and J. David Lockard. *Focus on Life Science*. Columbus, Ohio: C. E. Merrill, 1969.
William Keeton. *Elements of Biological Science*. New York: Norton, 1969.
Gordon Orians. *The Study of Life: An Introduction to Biology*. Boston: Allyn and Bacon, 1969.
A. M. Winchester. *Biology and Its Relation to Mankind*. New York: Van Nostrand Reinhold, 1969.
William Beaver and George Noland. *General Biology: The Science of Biology*. St. Louis: Mosby, 1970.
Relis Brown. *General Biology*. New York: McGraw-Hill, 1970.

Walter Thurber and Robert Kilburn. *Exploring Life Science*. Boston: Allyn and Bacon, 1970.
Alfred Elliott and Bruce Voeller. *Basic Biology*. New York: Appleton-Century-Crofts, 1970.
Coleman Goin and Olive Goin. *Man and the Natural World: An Introduction to Life Science*. New York: Macmillan, 1970.
Nancy Jessop. *Biosphere: A Study of Life*. Englewood Cliffs, N.J.: Prentice-Hall, 1970.
John Moore and Harold Slusher. *Biology: A Search for Order in Complexity*. Grand Rapids, Mich.: Zondervan, 1970.
V. Lawrence Parsegian. *Introduction to Natural Science Part Two: The Life Sciences*. New York: Academic Press, 1970.
James Casse and Vernon Stiers. *Biology: Observation and Concept*. New York: Macmillan, 1971.
Claude Villee and Vincent Dethier. *Biological Principles and Processes*. Philadelphia: Saunders, 1971.
James Ford and James Monroe. *Living Systems: Principles and Relationships*. San Francisco: Canfield Press, 1971.
Burton Guttman. *Biological Principles*. New York: W. A. Benjamin, 1971.
Jeffrey Baker and Garland Allen. *The Study of Biology*. Reading, Mass: Addison-Wesley, 1971.
Stanley Weinberg. *Biology: An Inquiry into the Nature of Life*. Boston: Allyn and Bacon, 1971.
Robert Korn and Ellen Korn. *Contemporary Perspectives in Biology*. New York: Wiley, 1971.
Edwin Phillips. *Basic Ideas in Biology*. New York: Macmillan, 1971.
Thomas Steyaert. *Life and Patterns of Order*. New York: McGraw-Hill, 1971.
Paul Weisz. *The Science of Biology*. New York: McGraw-Hill, 1971.
A. M. Winchester. *Modern Biological Principles*. New York: Van Nostrand Reinhold, 1971.
Paul Bailey and Kenneth Wagner. *An Introduction to Modern Biology*. Scranton: Intext Educational, 1972.
George Becker. *Introductory Concepts of Biology*. New York: Macmillan, 1972.
Stewart Brooks. *Basic Biology: A First Course*. St. Louis: Mosby, 1972.
William Etkin, Robert Devlin, and Thomas Bouffard. *A Biology of Human Concern*. Philadelphia: Lippincott, 1972.
Willis Johnson, Louis Delanney, Thomas Cole, and Austin Brooks. *Essentials of Biology*. New York: Holt, Rinehart and Winston, 1972.
Derry Koob and William Boggs. *The Nature of Life*. Reading, Mass: Addison-Wesley, 1972.
CRM Books. *Biology: An Appreciation of Life*. Del Mar, Calif.: CRM Books, 1972.
Gerard Tortora and Joseph Becker. *Life Science*. New York: Macmillan, 1972.
Claude Villee. *Biology*. Philadelphia: Saunders, 1972.
William Keeton. *Biological Science*. New York: Norton, 1972.
Neal Buffaloe and J. B. Throneberry. *Concepts of Biology: A Cultural Perspective*. Englewood Cliffs, N.J.: Prentice-Hall, 1973.
James Ebert. *Biology*. New York: Holt, Rinehart and Winston, 1973.
Paul Ehrlich, Richard Holm, and Michael Soulé. *Introductory Biology*. New York: McGraw-Hill, 1973.
William Keeton. *Elements of Biological Science*. New York: Norton, 1973.
Mary Clark. *Contemporary Biology: Concepts and Implications*. Philadelphia: Saunders, 1973.
Gairdner Moment and Helen Habermann. *Biology: A Full Spectrum*. London: Longman, 1973.

Alvin Nason and Robert L. Dehaan. *The Biological World.* New York: Wiley, 1973.
Michael Nicklanovich. *From Cell to Philosopher.* Englewood Cliffs, N.J.: Prentice-Hall, 1973.
David Nunnally and Burton Bogitsh. *Introductory Zoology.* New York: Wiley, 1973.
Gordon Orians. *The Study of Life: An Introduction to Biology.* Boston: Allyn and Bacon, 1973.
Roderick Suthers. *Biology: The Behavioral View.* Lexington, Mass: Xerox College, 1973.
Tully Turney. *An Introduction to Biology: The Subtle Science.* St. Louis: Mosby, 1973.
Aaron Wasserman. *Biology.* New York: Apple-Century-Crofts, 1973.
E. O. Wilson. *Life on Earth.* Stamford, Conn.: Sinauer, 1973.
James Ford and James Monroe. *Living Systems: Principles and Relationships.* San Francisco: Canfield Press, 1974.
Shelby Gerking. *Biological Systems.* Philadelphia: Saunders, 1974.
Coleman Goin and Olive Goin. *Man and the Natural World: An Introduction to Life Science.* New York: Macmillan, 1974.
Gideon Nelson and Gerald Robinson. *Fundamental Concepts of Biology.* New York: Wiley, 1974.
Grover Stephens and Barbara North. *Biology.* New York: Wiley, 1974.
Stanley Weinberg. *Biology: An Inquiry into the Nature of Life.* Boston: Allyn and Bacon, 1974.
Helena Curtis. *Biology.* New York: Worth, 1975.
Mahlon Kelly and John McGrath. *Biology: Evolution and Adaptation to the Environment.* Boston: Houston Mifflin, 1975.
George Noland and William Beaver. *General Biology.* St. Louis: Mosby, 1975.
B. S. Becket. *Biology: A Modern Introduction.* Oxford: Oxford University Press, 1976.
Paul Ehrlich, Richard Holm, and Irene Brown. *Biology and Society.* New York: McGraw-Hill, 1976.
Richard Goldsby. *Biology.* New York: Harper and Row, 1976.
William Keeton. *Biological Science.* New York: Norton, 1976.
Theodore Lane. *Life: The Individual and the Species.* St. Louis: C. V. Mosby, 1976.
Uldis Roze. *The Living Earth: An Introduction to Biology.* New York: Crowell, 1976.
Claude Villee and Vincent Dethier. *Biological Principles and Processes.* Philadelphia: Saunders, 1976.
James Ford and James Monroe. *Living Systems: Principles and Relationships.* San Francisco: Canfield Press, 1977.
Claude Villee. *Biology: The Human Approach.* Philadelphia: Saunders, 1977.
Edward Kormondy. *Biology: The Integrity of Organisms.* Belmont, Calif.: Wadsworth, 1977.
Gairdner B. Moment and Helen Habermann. *Mainstreams of Biology.* Baltimore: William and Wilkins, 1977.
Stephen Wolfe. *Biology: The Foundations.* Belmont, Calif.: Wadsworth, 1977.
Gerald Stine. *Biosocial Genetics: Human Heredity and Social Issues.* New York: Macmillan, 1977.
Jeffrey Baker and Garland Allen. *The Study of Biology.* Reading, Mass: Addison-Wesley, 1978.
Donald Farish. *The Human Perspective.* New York: Harper and Row, 1978.
Garrett Hardin and Carl Bajema. *Biology: Its Principles and Implications.* San Francisco: W. H. Freeman, 1978.
Charles Levy. *Elements of Biology.* Reading, Mass: Addison-Wesley, 1978.

Jean Macqueen and Ted Hanees. *The Living World: Exploring Modern Biology*. Englewood Cliffs, N.J.: Prentice-Hall, 1978.
Sam Singer and Henry Hilgard. *The Biology of People*. San Francisco: W. H. Freeman, 1978.
Cecie Starr and Ralph Taggart. *Biology: The Unity and Diversity of Life*. Belmont, Calif.: Wadsworth, 1978.
Mary Clark. *Contemporary Biology*. Philadelphia: W. B. Saunders, 1979.
William Davis and Eldra Solomon. *The World of Biology*. New York: McGraw-Hill, 1979.
Eldon Enger, Andrew Gibson, Richard Kromelink, Frederick Ross, and Bradley Smith. *Concepts in Biology*. Dubuque, Iowa: W. C. Brown, 1979.
Richard Goldsby. *Biology*. New York: Harper and Row, 1979.
George Noland. *General Biology*. St. Louis: Mosby, 1979.
Gideon Nelson. *Biological Principles with Human Perspectives*. New York: Wiley, 1980.
Pamela Camp and Karen Arms. *Exploring Biology*. Philadelphia: Saunders, 1981.
Salvador Luria, Stephen Jay Gould, and Sam Singer. *A View of Life*. Menlo Park, Calif.: Benjamin/Cummings, 1981.
Robert Wallace. *Biology: The World of Life*. Glenview, Ill.: Scott, Foresman, 1981.
Ruth Bernstein and Stephen Bernstein. *Biology: The Study of Life*. New York: Harcourt Brace Jovanovich, 1982.
Melissa Stanley and George Andrykovitch. *Living: An Interpretive Approach to Biology*. Reading, Mass: Addison-Wesley, 1982.
William Keeton and Carol McFadden. *Elements of Biological Science*. New York: Norton, 1983.
Knut Norstog and Andrew Meyerriecks. *Biology*. Columbus, Ohio: Merrill, 1983.
William Purves and Gordon Orians. *Life: The Science of Biology*. Sunderland, Mass: Sinauer, 1983.
Stephen Wolfe. *Biology: The Foundations*. Belmont, Calif.: Wadsworth, 1983.
Thomas Easton and Carl Rischer. *Bioscope*. Columbus, Ohio: C. E. Merrill, 1984.
Robert Wallace, Jack King, and Gerald Sanders. *Biosphere: The Realm of Life*. Glenview, Ill.: Scott, Foresman, 1984.
Eldon Gardner and Peter Snustad. *Principles of Genetics*. New York: Wiley, 1984.
Kenneth Johnson, David Rayle, and Hale Wedberg. *Biology: An Introduction*. Menlo Park, Calif.: Benjamin/Cummings, 1984.
Jack Ward and Howard Hetzel. *Biology: Today and Tomorrow*. St. Paul: West, 1984.
Pamela Camp and Karen Arms. *Exploring Biology*. Philadelphia: Saunders Collage, 1984.
Claude Villee, Eldra Solomon, and P. William Davis. *Biology*. Philadelphia: Saunders, 1985.
Robert Wallace, Jack King, and Gerald Sanders. *Biology: The Science of Life*. Glenview, Ill.: Scott, Foresman, 1986.
Teresa Audesirk and Gerald Audesirk. *Biology: Life on Earth*. Upper Saddle River, N.J.: Prentice Hall, 1986.
William Keeton, James Gould, and Carol Grant Gould. *Biological Science*. New York: Norton, 1986.
Thomas Overmire. *The World of Biology*. New York: Wiley, 1986.
Cecie Starr and Ralph Taggart. *Biology: The Unity and Diversity of Life*. Belmont, Calif.: Wadsworth, 1987.
Gilbert Brum and Larry McKane. *Biology: Exploring Life*. New York: Wiley, 1989.
Eldon Enger. *Concepts in Biology*. Dubuque, Iowa: William C. Brown, 1991.
Neil Campbell. *Biology*. Redwood City, Calif.: Benjamin/Cummings, 1993.
Neil Campbell. *Biology*. Menlo Park, Calif.: Benjamin/Cummings, 1996.

William Purves, Gordan Orians, Craig Heller, and David Sadava. *Life: The Science of Biology*. Sunderland, Mass: Sinauer, 1998.

Teresas Audesirk and Gerald Audesirk. *Biology: Life on Earth*. Upper Saddle River, N.J.: Prentice Hall, 1999.

Burton Guttman. *Biology*. New York: W. A. Benjamin, 1999.

Neil Campbell, Lawrence Mitchell, and Jane Reece. *Biology: Concepts and Connections*. San Francisco: Benjamin Cummings, 2000.

Scott Freeman. *Biological Science*. Harlow, England: Prentice Hall, 2002.

Eldra Solomon, Linda Berg, and Diana Martin. *Biology*. Pacific Grove, Calif.: Brooks/Cole Thomson Learning, 2002.

William Purves, David Sadava, Gordan Orians, and Craig Heller. *Life: The Science of Biology*. Sunderland, Mass: Sinauer, 2004.

Scott Freeman. *Biological Science*. Upper Saddle River, N.J.: Pearson Prentice Hall, 2005.

Peter Raven and George Johnson. *Biology: The Living Environment*. Orlando: Holt, Rinehart and Winston, 2005.

Notes

INTRODUCTION — IN THE NAME OF PROGRESS

1. For histories of coerced sterilization in the United States and of the American eugenics movement, see Mark Haller, *Eugenics: Hereditarian Attitudes in American Thought* (New Brunswick, N.J.: Rutgers University Press, 1963); Donald Pickens, *Eugenics and the Progressives* (Nashville, Tenn.: Vanderbilt University Press, 1968); Kenneth Ludmerer, *Genetics and American Society: A Historical Appraisal* (Baltimore: Johns Hopkins University Press, 1972); Garland Allen, "Genetics, Eugenics and Class Struggle," *Genetics* 79 (1975): 29–45; Garland Allen, "Genetics, Eugenics and Society: Internalists and Externalists in Contemporary History of Science," *Social Studies of Science* 6 (1976): 105–122; Stephen Jay Gould, *The Mismeasure of Man* (New York: W. W. Norton, 1981); Paul A. Lombardo, "Eugenic Sterilization in Virginia: Aubrey Strode and the Case of *Buck v. Bell*" (Ph.D. diss., University of Virginia, 1982); Paul A. Lombardo, "Involuntary Sterilization in Virginia: From *Buck v. Bell* to *Poe v. Lynchburg*," *Developments in Mental Health Law* 3 (1983): 17–21; Daniel Kevles, *In the Name of Eugenics: Genetics and the Uses of Human Heredity* (New York: Knopf, 1985); Paul A. Lombardo, "Three Generations, No Imbeciles: New Light on *Buck v. Bell*," *N.Y.U. Law Review* 60 (1985): 30–62; Garland Allen, "The Eugenics Record Office at Cold Spring Harbor, 1910–1940: An Essay in Institutional History," *Osiris* 2 (1986): 225–264; Troy Duster, *Backdoor to Eugenics* (New York: Routledge, 1990); Philip Reilly, *The Surgical Solution: A History of Involuntary Sterilization in the United States* (Baltimore: Johns Hopkins University Press, 1991); Edward J. Larson, *Sex, Race, and Science: Eugenics in the Deep South* (Baltimore: Johns Hopkins University Press, 1995); Diane Paul, *Controlling Human Heredity: 1865 to the Present* (Atlantic Highlands, N.J.: Humanities Press, 1995); Marouf Arif Hasian, *The Rhetoric of Eugenics in Anglo-American Thought* (Athens: University of Georgia Press, 1996); Martin Pernick, *The Black Stork: Eugenics and the Death of "Defective" Babies in American Medicine and Motion Pictures Since 1915* (Oxford: Oxford University Press, 1996); Steven Selden, *Inheriting Shame: The Story of Eugenics and Racism in America* (New York: Teachers College Press, 1999); Nancy Gallagher, *Breeding Better Vermonters: The Eugenics Project in the Green Mountain State* (Hanover, N.H.: University Press of New England, 1999); Philip J. Pauly, *Biologists and the Promise of American Life* (Princeton, N.J.: Princeton University Press, 2000); Elof Axel Carlson, *The Unfit: History of a Bad Idea* (Cold Spring Harbor, N.Y.: Cold Spring Harbor

Press, 2001); Wendy Kline, *Building a Better Race: Gender, Sexuality, and Eugenics from the Turn of the Century to the Baby Boom* (Berkeley: University of California Press, 2001); Edwin Black, *War against the Weak: Eugenics and America's Campaign to Create a Master Race* (New York: Four Walls Eight Windows, 2003); Nancy Ordover, *American Eugenics: Race, Queer Anatomy, and the Science of Nationalism* (Minneapolis: University of Minnesota Press, 2003); Christine Rosen, *Preaching Eugenics: Religious Leaders and the American Eugenics Movement* (Oxford: Oxford University Press, 2004); Christina Cogdale, *Eugenic Design: Streamlining America in the 1930s* (Philadelphia: University of Pennsylvania Press, 2004); Alexandra Minna Stern, *Eugenic Nation: Faults and Frontiers of Better Breeding in Modern America* (Berkeley: University of California Press, 2005); Johanna Schoen, *Choice and Coercion: Birth Control, Sterilization, and Abortion in Public Health and Welfare* (Chapel Hill: University of North Carolina Press, 2005); Harry Bruinius, *Better for All the World: The Secret History of Forced Sterilization and American's Quest for Racial Purity* (New York: Knopf, 2006).

2. Richard Lynn, *Eugenics: A Reassessment* (Westport, Conn.: Praeger, 2001); Nicholas Agar, *Liberal Eugenics: In Defense of Human Enhancement* (Oxford: Blackwell, 2004); John Glad, *Future Human Evolution: Eugenics in the Twenty-first Century* (Schuylkill Haven, Pa.: Hermitage, 2006).

3. Harry Laughlin, *Eugenical Sterilization in the United States* (Chicago: Psychopathic Laboratory of the Municipal Court of Chicago, 1922), 446.

4. My discussion of the use of particular terminology was significantly influenced by a similar discussion in the preface of Jesse F. Ballenger, *Self, Senility, and Alzheimer's Disease in Modern America: A History* (Baltimore: Johns Hopkins University Press, 2006), xiii–xiv.

5. C. Sengoopta, "'Dr Steinach Coming to Make Old Young!': Sex Glands, Vasectomy and the Quest for Rejuvenation in the Roaring Twenties," *Endeavor* 27 (2003): 122–126.

6. For an exploration of the notion of coercion as it relates to sterilization, see Schoen, *Choice and Coercion.*

7. I refer here to the title of Daniel Kevles's book on the subject, *In the Name of Eugenics.*

8. "Sterilization of Criminals," *Fortune* (June 1937): 106.

9. Sam Howe Verhovek, "Texas Agrees to Surgery for a Molester," *New York Times*, April 5, 1996; Sam Howe Verhovek, "Texas Frees Child Molester Who Warns of New Crimes," *New York Times*, April 9, 1996; ABC News, "Texas to Monitor Child Molester with Ankle Bracelet; Lawyer Says His Client Was Castrated," May 4, 2005.

10. The ratio of less than 1 in every 5,800 was produced by dividing the 1,333 sterilizations by the 1942 California population of 7,735,000. The total annual sterilizations of 1,333 for 1942 was generated by subtracting the cumulative sterilizations in California to January 1, 1942, of 15,220 as stated in Reilly, *Surgical Solution*, 97, as well as in Edgar Schmiedeler, *Sterilization in the United States: Introduction, Legal Status, the Question of Heredity, the Moral Question, the Solution* (Washington, D.C.: Family Life Bureau, National Catholic Welfare Conference, 1943) from the cumulative sterilizations in California to January 1, 1943, of 16,553 as stated in Jonas Robitscher, *Eugenic Sterilization* (Springfield, Ill.: Thomas, 1973): 118–119. California's population of 7,735,000 for 1942 is stated in California Statistical Abstract, 2000, California Department of Finance. www.dof.ca.gov/html/fs_data/STAT-ABS/toc.htm.

11. Curt Stern, *Principles of Human Genetics* (San Francisco: W. H. Freeman, 1949): 508–509.

12. "Eugenic," *Oxford English Dictionary*, 2nd ed., 1989.

CHAPTER 1 — NIPPING THE PROBLEM IN THE BUD

1. Charles Rosenberg, *No Other Gods: On Science and American Social Thought* (Baltimore: Johns Hopkins University Press, 1975), 47; J. E. Chamberlin and Sander L. Gilman, eds., *Degeneration: The Dark Side of Progress* (New York: Columbia University Press, 1985); Steven Gelb, "Degeneracy Theory, Eugenics, and Family Studies," *Journal of the History of the Behavioral Sciences* 26 (1990): 242–246.

2. For a full discussion of Lincecum's Memorial, see Lois Wood Burkhalter, *Gideon Lincecum, 1793–1874: A Biography* (Austin: University of Texas Press, 1965), 91–103.

3. Quoted in Burkhalter, *Gideon Lincecum*, 95–96.

4. Quoted in ibid., 96.

5. Ibid., 101.

6. Orpheus Everts, *Asexualization as a Penalty for Crime and Reformation of Criminals: A Paper Read Before the Cincinnati Academy of Medicine, February 27, 1888* (Cincinnati: W. B. Carpenter, 1888).

7. C. B. Gilbert et al., "The Prevention of Conception," *Cincinnati Medical News* 19 (1890): 303–308, described and cited in Norman Himes, *Medical History of Contraception* (Baltimore: Williams and Wilkins, 1936), 298.

8. William A. Hammond, "A New Substitute for Capital Punishment and Means for Preventing the Propagation of Criminals," *New York Medical Examiner* 114 (1891–1892): 190–194.

9. Ferdinand Eugene Daniel, "Castration of Sexual Perverts," *Texas Medical Journal* 8 (1893): 255–271.

10. Ibid.; Ferdinand Eugene Daniel, "Should Insane Criminals or Sexual Perverts be Allowed to Procreate?" *Medico-Legal Journal* 114 (1893–1894): 275–292.

11. A. C. Ames, "A Plea for Castration, as a Punishment for Crime," *Omaha Clinic* 6 (1893–1894): 343–345.

12. F. L. Sim, "Asexualization for the Prevention of Crime and the Arrest of the Propagation of Criminals," *Medical Society of Tennessee* (1894): 100–114.

13. B. A. Arbogast, "Castration as a Remedy for Crime," *Denver Medical Times* 15 (1895): 55–58.

14. E. Stuver, "Asexualization for the Limitation of Disease and the Prevention and Punishment of Crime," *Transactions of the Colorado State Medical Society* (1895): 327–336.

15. A. C. Corr, "Emasculation and Ovariotomy as a Penalty for Crime and as a Reformatory Agency," *Medical Age* 13 (1895): 714–716.

16. Ira M. Rutkow, "Orificial Surgery," *Archives of Surgery* 136 (2001): 1088; Ira M. Rutkow, "Edwin Hartley Pratt and Orificial Surgery: Unorthodox Surgical Practice in Nineteenth Century United States," *Surgery* 114 (1993): 558–563.

17. Edwin Hartley Pratt, *Orificial Surgery and Its Application to the Treatment of Chronic Diseases* (Chicago: Halsey Brothers, 1887).

18. E. H. Pratt, "A Monthly Series of Articles Upon the Orificial Philosophy Written for the Journal of Orificial Surgery," *Journal of Orificial Surgery* 1 (1892): 22–27.

19. "Clippings and Comments," *Journal of Orificial Surgery* 1 (1892): 521.

20. For a complete examination of the origins and practices of orificial surgery, see B. E. Dawson, ed., *Orificial Surgery: Its Philosophy, Application, and Technique* (Chicago: Adams, 1912).

21. S. C. Gordon, "Hysteria and Its Relation to Disease of the Uterine Appendages," *Journal of the American Medical Association* 6 (1886): 561–567.

22. T. Spencer Wells, "Castration in Mental and Nervous Diseases: A Symposium," *American Journal of Medical Sciences* 92 (1886): 455–471.

23. Alfred Hegar, "Castration in Mental and Nervous Diseases: A Symposium," *American Journal of Medical Sciences* 92 (1886): 471–483.

24. Robert Battey, "Castration in Mental and Nervous Diseases: A Symposium," *American Journal of Medical Sciences* 92 (1886): 483–490.

25. T. Griswold Comstock, "Alice Mitchell of Memphis: 'Clippings and Comments,'" *Journal of Orificial Surgery* 1 (1892–1893): 207.

26. F. L. Sim, "Forensic Psychiatry: Alice Mitchell Adjudged Insane," *Memphis Medical Monthly* 12 (1892): 377–428; Daniel, "Castration of Sexual Perverts," 255–271. Both Sim's and Daniel's works are cited and discussed in Jonathan Ned Katz, ed., *Gay American History: Lesbians and Gay Men in the U.S.A.*, rev. ed. (New York: Penguin, 1992), 53–58, 135–137.

27. C. A. Kirkley, "Gynecological Observations in the Insane," *Journal of the American Medical Association* 19 (1892): 553–555.

28. Bible, New International Version, Genesis 39: 7–10.

29. *Herring v. State*, 119 Ga. 709, 46 S.E. 876; 1904.

30. *State v. Fenner*, 166 N.C. 247, 80 S.E. 970; 1914.

31. James Spence, "The Law of the Crime against Nature," *North Carolina Law Review* 32 (1953–1954): 312–324.

32. Charles H. Hughes, "An Emasculated Homo-Sexual: His Antecedent and Post-Operative Life," *Alienist and Neurologist* 35 (1914): 277–280.

33. Mark A. Largent, "'The Greatest Curse of the Race:' Eugenic Sterilization in Oregon, 1909–1983," *Oregon Historical Quarterly* 103 (2002): 188–209; Kevin Begos, "Castration: Files Suggest That Punishment Was Often the Aim," http://against.theirwill.journalnow.com.

34. R. L. Steiner, "Eugenics in Oregon," *Northwest Medicine* 26 (December 1927): 594–597.

35. Karl M. Bowman and Bernice Engle, "The Problem of Homosexuality," *Journal of Social Hygiene* 39 (1953): 7–9, 10–11.

36. "Removal of the Ovaries as a Therapeutic Measure in Public Institutions for the Insane," *Journal of the American Medical Association* 20 (1893): 135–137.

37. Mark Millikin, "The Proposed Castration of Criminals and Sexual Perverts," *Cincinnati Lancet-Clinic* 33 (1894): 185–190.

38. Dean Collins, "Children of Sorrow: A History of the Mentally Retarded in Kansas," *Bulletin of the History of Medicine* 39 (1965): 59–61.

39. Frederick D. Seaton, "The Long Road toward 'The Right Thing to Do,'" *Kansas History: A Journal of the Central Plains* 27 (2004–2005): 253.

40. F. C. Cave, "Report of Sterilization in the Kansas State Home for Feeble-Minded," *Journal of Psycho-Asthenics* 15 (1910–1911): 123–125.

41. R. E. M'Vey, "Crime—Its Physiology and Pathogenesis. How Can Medical Men Aid in Its Prevention?" *Kansas Medical Journal* 2 (1890): 499–504.

42. Bernard Douglass Eastman, "Can Society Successfully Organize to Prevent Over-Production of Defectives and Criminals?" *Topeka Medical Journal* 9 (1897): 351–355.

43. "Mutilation," *Winfield Daily Courier*, August 24, 1894.

44. "The Winfield Case," *Kansas Medical Journal* (September 8, 1894); reprinted in "Emasculation of Masturbators—Is It Justifiable?" *Texas Medical Journal* 109 (1894): 243–244.

45. Henry Roby, "Family Doctor," *Kansas Farmer*; reprinted in "Emasculation of Masturbators," 239–241.

46. "Emasculation of Masturbators," 239.

47. Cave, "Report of Sterilization," 123–125.

48. G. Frank Lydston, *Addresses and Essays* (Louisville: Renz and Henry, 1892); G. Frank Lydston, "Sexual Perversion, Satyriasis, and Nymphomania," *Medical and Surgical Reporter* 61 (1889): 253.

49. G. Frank Lydston and Hunter McGuire, *Sexual Crimes among the Southern Negroes* (Louisville: Renz and Henry, 1893).

50. G. Frank Lydston, *The Diseases of Society: The Vice and Crime Problem* (Philadelphia: J. B. Lippincott, 1904), 41–79.

51. Ibid., 48.

52. Ibid., 69.

53. Ibid., 557–558.

54. Ibid., 562.

55. Ibid., 564–565.

56. See William M. Kantor, "Beginnings of Sterilization in America: An Interview with Dr. Harry C. Sharp, Who Performed the First Operation Nearly Forty Years Ago," *Journal of Heredity* 28 (1937): 375; and Harry C. Sharp, "The Severing of the Vasa Deferentia and Its Relation to the Neuropsychopathic Constitution," *New York Medical Journal* (1902): 413.

57. "Whipping and Castration as Punishments for Crime," *Yale Law Journal* 8 (1899): 371–386.

58. Jesse Ewell, "A Plea for Castration to Prevent Criminal Assault," *Virginia Medical Monthly* (January 11, 1907): 463–464.

59. For example, see "Eugenic Sterilization in Indiana," *Indiana Law Journal* 38 (1962–1963): 276.

60. A. Cooper, *Observations on the Structure and Diseases of the Testis* (London: McDowall, 1830); M. J. Drake, I. W. Mills, and D. Cranston, "On the Chequered History of the Vasectomy," *British Journal of Urology* 84 (1999): 475.

61. A. J. Ochsner, "Surgical Treatment of Habitual Criminals," *Journal of the American Medical Association* 32 (1899): 867–868.

62. Reilly, *Surgical Solution*, 30.

63. Ochner, "Surgical Treatment of Habitual Criminals," 867–868.

64. Kantor, "Sterilization in America," 374–376; Angela Gugliotta, "'Dr. Sharp with His Little Knife:' Therapeutic and Punitive Origins of Eugenic Vasectomy—Indiana, 1892–1921," *Journal of the History of Medicine* 53 (1998): 371–406.

65. Kantor, "Sterilization in America," 374–376.

66. Sharp, "Severing of the Vasa Deferentia," 411.

67. Harry C. Sharp, "Vasectomy as a Means of Preventing Procreation in Defectives," *Journal of the American Medical Association* 53 (1909): 1897–1902.

68. Harry C. Sharp, "The Indiana Idea of Human Sterilization," *Southern California Practitioner* 24 (1909): 549–551.

69. Leila Zenderland, *Measuring Minds: Henry Herbert Goddard and the Origins of American Intelligence Testing* (Cambridge: Cambridge University Press, 1998), 150.

70. Martin Barr, *Mental Defectives: Their History, Treatment, and Training* (Philadelphia: Blakiston, 1904), 95–97, 108; cited in Zenderland, *Measuring Minds*, 397.

71. By 1913, compulsory sterilization laws were passed in Indiana, Washington, California, Connecticut, Iowa, New Jersey, Nevada, New York, North Dakota, Michigan, Kansas, and Wisconsin. Their cumulative population, based on the 1910 census, was 26,516,741, and the national population was 91,641,195. Governors in Pennsylvania, Oregon, Vermont, and Nebraska vetoed compulsory sterilization bills.

72. William T. Belfield, "The Sterilization of Criminals and Other Defectives by Vasectomy," *Journal of the New Mexico Medical Society* (July 1909): 21–25; *State v. Feilen*, 70 Wash. 65; 126 P. 75; 1912; "Constitutional Law: State Sterilization Law Not Contrary to the Fourteenth Amendment Giving Due Process and Equal Protection of the Law," *Marquette Law Review* 12 (1927–1928): 73–76.

73. "Race Suicide for Social Parasites," *Journal of the American Medical Association* 50 (1908): 55–56.

74. "Vasectomy for Confirmed Criminals and Defectives," *Journal of the American Medical Association* 53 (1909): 1114.

75. Belfield, "Sterilization of Criminals," 21–25; "Editorial Comment," *Journal of the New Mexico Medical Society* (1909): 24–25.

76. William H. Houser, "To Secure Immunity from Crime," *Ellingwood's Therapeutist* 2 (May 15, 1908): 19–21.

77. J. Ewing Mears, "Asexualization as a Remedial Measure in the Relief of Certain Forms of Mental, Moral and Physical Degeneration," *Boston Medical and Surgical Journal* 161 (1909): 584–586; "Race Suicide," 55.

78. Henri Bogart, "Sterilization of the Unfit—The Indiana Plan," *Medical Herald* 30 (1911): 81–84.

79. These quotes were taken from James B. O'Hara and T. Howland Sanks, "Eugenic Sterilization," *Georgetown Law Journal* 45 (1956): 21. They are representative of statements commonly found in both the primary and secondary source material describing the relative ease and freedom from side effects common to vasectomies and tubal ligations.

80. Sharp, "Severing of the Vasa Deferentia," 411.

81. Everts, *Asexualization*; Hammond, "New Substitute for Capital Punishment"; Ames, "Plea for Castration"; Sim, "Asexualization."

82. Robert Reid Rentoul, *Proposed Sterilization of Certain Mental and Physical Degenerates: An Appeal to Asylum Managers and Others* (London: Walter Scott, 1903).

83. Robert Rentoul, "Proposed Sterilization of Certain Mental Degenerates," *American Journal of Sociology* 12 (1906): 319–327.

84. Robert Reid Rentoul, *Race Culture: Or, Race Suicide? (A Plea for the Unborn)* (London: Walter Scott, 1906).

85. William A. Challener Jr., "The Law of Sexual Sterilization in Pennsylvania," *Dickinson Law Review* 57 (1953): 298.

86. Martin W. Barr, "President's Annual Address," *Journal of Psycho-Asthenics* 2 (1897): 1–13.

87. Martin W. Barr, "Some Notes on Asexualization: With a Report of Eighteen Cases," *Journal of Nervous and Mental Diseases* 51 (1920): 231–241.

88. S. D. Risley, "Is Asexualization Ever Justifiable in the Case of Imbecile Children," *Journal of Psycho-Asthenics* 9 (1905): 93–98.

89. "Editorial: Sterilization of the Unfit," *Journal of Psycho-Asthenics* 9 (1905): 127–128; Christian Kellar, "Asexualization—Attitude of Europeans," *Journal of Psycho-Asthenics* 9 (1905): 128–129; Martin W. Barr, "Results of Sterilization," *Journal of Psycho-Asthenics* 9 (1905): 129.

90. Edwin A. Down, *The Sterilization of Degenerates* (Hartford, Conn., 1910).

91. Thorsten Sellin, "Hastings Hornell Hart," *Journal of Criminal Law and Criminology* 23 (1932): 166.

92. Hastings H. Hart, *Sterilization as a Practical Measure: A Paper Read before the American Prison Association at Baltimore November 12th, 1912* (New York City: Department of

Child Helping of the Russell Sage Foundation, 1913); Hastings H. Hart, *The Extinction of the Defective Delinquent: A Working Program: A Paper Read before the American Prison Association at Baltimore November 12, 1912* (New York City: Department of Child Helping of the Russell Sage Foundation, 1913); Hastings H. Hart, "A Working Program for the Extinction of the Defective Delinquent," *Survey* 30 (1913): 277–279.

CHAPTER 2 — EUGENICS AND THE PROFESSIONALIZATION OF AMERICAN BIOLOGY

1. E. Carleton MacDowell, "Charles Benedict Davenport: A Study of Conflicting Influences," *Bios* 17 (1946).
2. Ibid., 15.
3. Charles Davenport, "Zoology of the Twentieth Century," *Science* 14 (1901): 315–324.
4. David Starr Jordan to Gertrude Crotty Davenport, December 12, 1905, Charles Benedict Davenport Papers, American Philosophical Society, Philadelphia. Jordan suggested that she look at his *Footnotes to Evolution*, in which he extracted information from Oscar McCulloch's study on the "Tribe of Ishmael." See David Starr Jordan, *Footnotes to Evolution* (New York: Appleton, 1898).
5. Gertrude Crotty Davenport to B. K. Bruce, February 24, 1907, Davenport Papers, American Philosophical Society.
6. Charles Davenport to the Trustees of the Carnegie Institution, March 5, 1903, SEE Application to CIW, Carnegie Institution of Washington Archives. A thorough examination of the creation of the Cold Spring Harbor Station for Experimental Evolution and Davenport's role is found in Pamela Mack, "The Early Years of Cold Spring Harbor Station for Experimental Evolution," unpublished manuscript (1979), American Philosophical Society.
7. Allen, "Eugenics Record Office," 229.
8. Charles Davenport to Alexander Agassiz, May 30, 1902, SEE Application to CIW, Carnegie Institution of Washington Archives.
9. Charles Davenport to Frank Billings, May 3, 1903, SEE Application to CIW, Carnegie Institution of Washington Archives.
10. Ibid.
11. Kevles, *In the Name of Eugenics*, 48.
12. Pauly, *Biologists and the Promise of American Life*, 220.
13. Allen, "Eugenics Record Office"; Charles Davenport, "Mendel's Law of Dichotomy in Hybrids," *Biological Bulletin* 2 (1901): 307–310.
14. MacDowell, "Charles Benedict Davenport," 26.
15. Charles Davenport, *Inheritance in Canaries* (Washington, D.C.: Carnegie Institution of Washington, 1908), 5, 23; Charles Davenport, *Inheritance in Poultry* (Washington, D.C.: Carnegie Institution of Washington, 1906).
16. A. Rudolf Galloway, "Canary Breeding: A Partial Analysis of Records from 1891–1909," *Biometrika* 7 (1909): 1–42.
17. Charles Davenport, "Dr. Galloway's 'Canary Breeding,'" *Biometrika* 7 (1910): 398–400.
18. See Charles Davenport to Major Leonard Darwin, April 9, 1927, Davenport Papers, American Philosophical Society; Theodore M. Porter, *Karl Pearson: The Scientific Life in a Statistical Age* (Princeton, N.J.: Princeton University Press, 2004), 270.
19. A. Rudolf Galloway, "Canary Breeding: A Rejoinder to C. B. Davenport," *Biometrika* 7 (1910): 401–403; David Heron: "Inheritance in Canaries: A Study in Mendelism," *Biometrika* 7 (1910): 403–410. For the fight between Davenport and Heron in the *New York*

Times, see "English Expert Attacks American Eugenics Work," *New York Times*, November 9, 1913; "American Work Strongly Defended: Dr. Davenport, Who Is the Centre of Criticism, Replies to Dr. Heron's Attacks: English Attack on Our Eugenics," *New York Times*, November 9, 1913.

20. Hamish G. Spencer and Diane B. Paul, "The Failure of a Scientific Critique: David Heron, Karl Pearson, and Mendelian Eugenics," *British Journal of the History of Science* 31 (1998): 441–452.

21. Charles Davenport and Mary Theresa Scudder, *Naval Officers: Their Heredity and Development* (Washington, D.C.: Carnegie Institution of Washington, 1919); Charles Davenport, "On Utilizing the Facts of Juvenile Promise and Family History in Awarding Naval Commissions to Untried Men," *Proceedings of the National Academy of Science of the United States of America* 3 (1917): 404–409.

22. L. L. Bernard, "Review of Davenport's Naval Officers: Their Heredity and Development," *American Journal of Sociology* 25 (1919): 241–242.

23. "Proceedings of the First Meeting of the American Breeders Association held at St. Louis, Missouri, December 29 and 30, 1903," *American Breeders Association Proceedings* 1 (1905): 9–15.

24. See "The New Magazine," *American Breeders Magazine* 1 (1910): 61.

25. Ibid.

26. "These Are Times of Scientific Ideals," *American Breeders Magazine* 4 (1913): 61.

27. Charles Davenport, "The Relation of the Association to Pure Research," *American Breeders Magazine* 1 (1910): 66–67.

28. "Scientific Ideals," 61.

29. David Starr Jordan, "Report of the Committee on Eugenics," *American Breeders Association Proceedings* 45 (1908): 201–202; "Committee on Eugenics," *American Breeders Association Proceedings* 2 (1906): 11.

30. "Committee on Eugenics," 11.

31. "Minutes," *American Breeders Association: Proceedings of the Meeting Held at Columbus, Ohio, January 15–18, 1907, III* (Washington, D.C.: American Breeders Association, 1907), 137.

32. Jordan, "Report," 202.

33. Charles Davenport to David Starr Jordan, January 27, 1910, Davenport Papers, American Philosophical Society; David Starr Jordan to Charles Davenport, February 1, 1910, Davenport Papers, American Philosophical Society.

34. Willet M. Hays, "Proposed Change in the Constitution of the American Breeders Association," *American Breeders Magazine* 1 (1910): 75. See also "The Section Organization," *American Breeders Magazine* 1 (1910): 143; and "The New Eugenics Section," *American Breeders Magazine* 1 (1910): 153.

35. "The Field of Eugenics," *American Breeders Magazine* 2 (1911): 141.

36. Charles Davenport, "Eugenics, a Subject for Investigation Rather Than Instruction," *American Breeders Magazine* 1 (1910): 68.

37. "The Movement to Study Eugenics," *American Breeders Magazine* 1 (1910): 143; "The Pedagogics of Eugenics," *American Breeders Magazine* 3 (1912): 224.

38. "Preachers and Eugenics," *American Breeders Magazine* 4 (1913): 63.

39. "Pedagogics of Eugenics," 222.

40. Vernon Lyman Kellogg, "Man and the Laws of Heredity," *American Breeders Association Proceedings* 5 (1909): 242.

41. Frederick Adams Woods, "Some Desiderata in the Science of Eugenics," *American Breeders Association Proceedings* 5 (1909): 248.

42. Charles Woodruff, "Prevention of Degeneration the Only Practical Eugenics," *American Breeders Association Proceedings* 3 (1907): 247–252.

43. Ibid., 248.

44. "Pedagogics of Eugenics," 222.

45. Alexander Graham Bell, "A Few Thoughts Concerning Eugenics," *American Breeders Association Proceedings* 4 (1908): 209.

46. Alexander Graham Bell, "Eugenics," *American Breeders Association Proceedings* 5 (1909): 218–220.

47. Charles Davenport, "Report of Committee on Eugenics," *American Breeders Magazine* 1 (1910): 126–129.

48. Woodruff, "Prevention of Degeneration the Only Practical Eugenics," 251.

49. Bell, "Few Thoughts Concerning Eugenics," 208. A similar sentiment is expressed in Charles E. Woodruff, "Climate and Eugenics," *American Breeders Magazine* 1 (1910): 183–185.

50. Roswell H. Johnson, "The Direct Action of the Environment," *American Breeders Association Proceedings* 5 (1909): 228.

51. See, for example, Maynard M. Metcalf, "Eugenics and Euthenics," *Popular Science Monthly* 84 (1914): 383–389; and Leon J. Cole, "The Relation of Eugenics to Euthenics," *Popular Science Monthly* 81 (1912): 475–482.

52. Barbara Kimmelman, "The American Breeders Association: Genetics and Eugenics in an Agricultural Context," *Social Studies of Science* 13 (1983): 189. See also Barbara Kimmelman, "A Progressive Era Discipline: Genetics at American Agricultural Colleges and Experiment Stations, 1900–1920" (Ph.D. diss., University of Pennsylvania, 1987).

53. See "Secretary's Report," *American Breeders Association Proceedings* 8 (1912): 324–335.

54. Kimmelman, "American Breeders Association," 183–194.

55. "Breeders Association Will Change Its Name," *American Breeders Magazine* 4 (1913): 177.

56. George Kennan, *E. H. Harriman* (Boston: Houghton-Mifflin, 1922).

57. Allen, "Eugenics Record Office," 235.

58. MacDowell, "Charles Benedict Davenport," 29.

59. Harry Laughlin to Charles Davenport, February 25, 1907, Davenport Papers, American Philosophical Society.

60. Harry Laughlin to Charles Davenport, March 30, 1908, Davenport Papers, American Philosophical Society.

61. Harry Laughlin to Charles Davenport, February 24, 1909, Davenport Papers, American Philosophical Society.

62. See Frances Hassencahl, "Harry H. Laughlin, 'Expert Eugenical Agent' for the House Committee on Immigration and Naturalization" (Ph.D. diss., Case Western Reserve, 1970).

63. Allen, "Eugenics Record Office," 226.

64. Harry H. Laughlin, "The Eugenics Record Office at the End of Twenty-seven Months' Work," *Report of the Eugenics Record Office* 1 (1912).

65. Laughlin, *Eugenical Sterilization.*

66. Charles Davenport, *Heredity in Relation to Eugenics* (New York: Henry Holt, 1911), iii, 1, 4.

67. Ibid., 80–82, 254, 255.

68. Ibid., 255–259.

69. Ibid., 256–259.

70. *Osborn v. Thomson, Andres, and Wansboro, Composing the Examiners of Feeble-Minded Criminals and Other Defectives*, 185 A.D. 902; 171 N.Y.S. 1094; 1918.

71. Henry Herbert Goddard, *The Kallikak Family: A Study in the Heredity of Feeble-Mindedness* (New York: Macmillan, 1922), 101, 105, 108, 113. See also Zenderland, *Measuring Minds*, 182.

72. Allen, "Eugenics Record Office," 254.

73. MacDowell, "Charles Davenport"; Oscar Riddle, "Charles Benedict Davenport," *Biographical Memoirs of the National Academy of Sciences* 25 (1949): 75–110.

CHAPTER 3 — THE LEGISLATIVE SOLUTION

1. Albert Gray, "Notes on the State Legislation of America in 1895," *Journal of the Society of Comparative Legislation* 1 (1896–1897): 233.

2. "Legislation against the Propagation of Disease," *West Virginia Bar* 3 (1896): 160–161.

3. Jessie Spaulding Smith, "Marriage, Sterilization and Commitment Laws Aimed at Decreasing Mental Deficiency," *Journal of the American Institute of Criminal Law and Criminology* 5 (1914): 364–370; Edward W. Spencer, "Some Phases of Marriage Law and Legislation from a Sanitary and Eugenic Standpoint," *Yale Law Journal* 25 (1915): 58–73.

4. John P. Jackson, *Science for Segregation: Race, Law and the Case against* Brown v. Board of Education (New York: New York University Press, 2005), 30–31.

5. R. H. G., "State Laws Regulating Marriage of the Unfit," *Journal of the American Institute of Criminal Law and Criminology* 4 (1913): 423–425; Smith, "Marriage, Sterilization and Commitment Laws," 364–370.

6. "Notes," *Albany Law Journal* 55 (1897): 375–376.

7. Laughlin, *Eugenical Sterilization*, 35.

8. "Notes," 375.

9. "Our Supplement," *American Lawyer* 5 (1897): 299.

10. W. R. Edgar, "Asexualization of Criminals and Degenerates," *Michigan Law Journal* 6 (1897): 291–294.

11. William M. Donald, "Asexualization of Criminals and Degenerates," *Michigan Law Journal* 6 (1897): 294–297; David Inglis, "Asexualization of Criminals and Degenerates," *Michigan Law Journal* 6 (1897): 298–300; J. J. Mulheron, "Asexualization of Criminals and Degenerates," *Michigan Law Journal* 6 (1897): 301.

12. Max Simon Nordau, *Degeneration* (New York: D. Appleton, 1895). Donald referred to "Lombrosi," apparently a misspelling of Lombroso.

13. Donald, "Asexualization," 296; Inglis, "Asexualization," 300.

14. Donald, "Asexualization," 296.

15. Inglis, "Asexualization," 300.

16. Edgar, "Asexualization," 291, 292.

17. Donald, "Asexualization," 296.

18. John E. Clark, "Asexualization of Criminals and Degenerates," *Michigan Law Journal* 6 (1897): 301–305; Charles W. Hitchcock, "Asexualization of Criminals and Degenerates," *Michigan Law Journal* 6 (1897): 305–307; Charles S. Morley, "Asexualization of Criminals and Degenerates," *Michigan Law Journal* 6 (1897): 307–310; Orville W. Owen, "Asexualization of Criminals and Degenerates," *Michigan Law Journal* 6 (1897): 311–312; Clarence Lightner, "Asexualization of Criminals and Degenerates," *Michigan Law Journal* 6 (1897): 312–315.

19. Clark, "Asexualization," 303. Just as Donald had misspelled Lombroso's name, Clark spelled Galton's name as Gaulton.

20. Morley, "Asexualization," 310.

21. Hitchcock, "Asexualization," 307.

22. Lightner, "Asexualization," 312–315.

23. Edgar, "Asexualization," 292.

24. Michigan Legislature, *Journal of the House of Representatives of the State of Michigan, 1897* (Detroit: Morse and Bagg, 1897), 2140.

25. David Johnson, "Out of the Ashes," *Michigan History Magazine* (January–February 2001): 22–27. The fire was set by a nineteen-year-old state employee who hoped it would be quickly contained so that he could admit to it, be placed on probation, and avoid being drafted and sent to Korea.

26. Challener, "Law of Sexual Sterilization," 298.

27. Isaac Kerlin, "President's Address," *Proceedings of the Association of Medical Officers of American Institutions for Idiotic and Feeble-Minded Persons* (1892): 274.

28. Vetoes by the Governor of Bills Passed by the Legislature, Pennsylvania, Session of 1905, 26.

29. Edward Leroy Van Roden, "Sterilization of Abnormal Persons as Punishment for and Prevention of Crimes," *Temple Law Quarterly* 23 (1949): 106.

30. Largent, "Greatest Curse."

31. Indiana Acts, 1907, Chapter 215.

32. *Williams et al. v. Smith*, 190 Ind. 526; 131 N.E. 2; 1921.

33. Indiana Acts 1927, Chapter 241; Frank T. Lindman and Donald M. McIntyre, *The Mentally Disabled and the Law* (Chicago: University of Chicago Press, 1961), 190; "Eugenic Sterilization in Indiana," 275–288.

34. Indiana Acts 1931, Chapter 50; Indiana Acts 1935, Chapter 12.

35. "Indiana Sterilization Bill Voted," *New York Times*, March 9, 1937.

36. C. O. McCormick, "Is the Indiana 1935 Sterilization of the Insane Act Functioning?" *Journal of the Indiana State Medical Association* 42 (1949): 919–920.

37. Robitscher, *Eugenic Sterilization*, 118–119.

38. "Eugenic Sterilization in Indiana."

39. "Vasectomy for Confirmed Criminals and Defectives," 1114.

40. "Draft of Sterilization Law for Illinois," *Journal of the American Institute of Criminal Law and Criminology* 7 (1916): 611–614.

41. E. R. K., "Sterilization of Habitual Criminals and Feeble-Minded Persons," *Illinois Law Review* 5 (1910–1911): 578.

42. "Draft of Sterilization Law for Illinois," 611–614.

43. Pernick, *Black Stork*, 4.

44. W. F. Gray, "On Legislation for Sterilization," *Journal of the American Institute of Criminal Law and Criminology* 7 (1916): 591–592.

45. "Would Sterilize the 'Socially Inadequate,'" *New York Times*, January 4, 1923.

46. "Description of Harry Olson Papers," 1906–1940, Northwestern University Archives, Evanston, Ill. See also Michael Willrich, "The Two Percent Solution: Eugenic Jurisprudence and the Socialization of American Law," *Law and History Review* 16 (1998): 63–111.

47. Laughlin, *Eugenical Sterilization*; Harry Laughlin, "The Fundamental Biological and Mathematical Principles Underlying Chromosomal Descent and Recombination in Human Heredity," in *Research Studies of Crime as Related to Heredity* (Chicago: Municipal Court of Chicago, 1925), 74–82.

48. Abraham Myerson, James B. Ayer, Tracy Putnam, Clyde Keeler, and Leo Alexander, *Eugenical Sterilization: A Reorientation of the Problem* (New York: Macmillan, 1936).

49. Marie E. Kopp, "Surgical Treatment as Sex Crime Prevention Measure," *Journal of Criminal Law and Criminology* 29 (1938): 697.

50. Kerlin, "President's Address," 278.

51. Laughlin, *Eugenical Sterilization*, 35–40.

52. Burkhalter, *Gideon Lincecum*, 101.

53. "Emasculation for Criminal Assaults and for Incest," *Texas Medical Journal* 22 (1907): 347.

54. *Ohio Senate Journal, 1927*, 191, 202, 320–321; *Ohio House of Representatives Journal, 1927*, 472, 522, 627.

55. *In re Simpson*, 180 N.E. 2d 206; 1962.

56. See Pickens, *Eugenics*; and Rudolph J. Vecoli, "Sterilization: A Progressive Measure?" *Wisconsin Magazine of History* 43 (1960): 190–202.

57. Jay F. Arnold, "A Sterilization Law for Kentucky—Its Constitutionality," *Kentucky Law Journal* 24 (1935–1936): 220–229; Jay F. Arnold, "Sociological Expediency of Sterilization Statute," *Kentucky Law Journal* 25 (1936–1937): 186–192.

58. "Notes and Legislation," *Iowa Law Review* 35 (1949–1950): 251–304. The article cited an interview with Dr. W. R. Miller, medical director of the State Psychopathic Hospital, Iowa City, Iowa, and member of the Board of Eugenics.

59. *Montana Code Annual*, §§ 53–23–101-53–23–105 (1969).

60. *Utah Code Annual*, § 64–10–8 (1975).

61. Stanley Powell Davies, *Social Control of the Mentally Deficient* (New York: Thomas Y. Crowell, 1930), 128. See also Larson, *Sex, Race and Science.*

62. California Statutes, 1909, Chapter 720.

63. California Statutes, 1913, Chapter 363.

64. California Statutes, 1917, Chapter 489.

65. Carl Ingram, "State Issues Apology for Policy of Sterilization," *Los Angeles Times*, March 12, 2003.

66. Stern, *Eugenic Nation.*

67. Laws of North Dakota 1913, Chapter 56.

68. Joel D. Hunter, "Sterilization of Criminals (Report of Committee H of the Institute)," *Journal of the American Institute of Criminal Law and Criminology* 5 (1914): 525.

69. Laws of North Dakota 1927, Chapter 263.

70. Duane R. Nedrud, "Sterilization—Scope of the State's Power to Use Sterilization on Mental Defectives and Criminals—Operation of North Dakota Statute," *North Dakota Bar Briefs* 26 (1950): 190.

71. *Miranda v. Arizona*, 396 U.S. 868; 90 S. Ct. 140; 24 L. Ed. 2d 122; 1969.

72. Robitscher, *Eugenic Sterilization*, 118–119. Julius Paul, "Population 'Quality' and 'Fitness for Parenthood' in the Light of State Eugenic Sterilization Experience," *Population Studies* 21 (1967): 297.

73. Session Laws of Washington, 1909, Chapter 249, Section 35.

74. *State v. Feilen.*

75. Session Laws of Washington, 1921, Chapter 53.

76. Christine Manganaro, "Eugenicist as Patient Advocate: Therapeutic Sterilization in Washington State" (Honor's thesis: University of Puget Sound, 2003). Manganaro argues that the sterilizations Keller performed on inmates of the Western State Hospital at Fort Steilacoom were motivated by therapeutic factors, rather than punitive or eugenic.

77. *In the Matter of the Sterilization of Hollis Hendrickson*, 123 P.2d 322; 1942.

78. Washington Revised Annual Code, 9.92.100.

79. Nevada Statutes, 1911, Chapter 28.

80. Laughlin, *Eugenical Sterilization*, 20–21.

81. *Mickle v. Henrichs, Warden of Nevada State Prison, et al.*, 262 F. 687; 1918.

82. Iowa Laws, 1913, Chapter 187.

83. Laughlin, *Eugenical Sterilization*, 67; Iowa Laws, 1911, Chapter 129.

84. *David v. Berry et al.*, 216 F. 413; 1914.

85. *Berry et al. Constituting the Board of Parole of Iowa, et al. v. Davis*, 242 U.S. 468; 37 S. Ct. 208; 61 L. Ed. 441; 1917.

86. Iowa Laws, 1915, Chapter 202; Iowa Laws, 1946, Chapter 145.

87. "Notes and Legislation," 251–304.

88. Ibid., 264.

89. Ibid., 261; Robitscher, *Eugenic Sterilization*, 118–119.

90. New Jersey Acts of the Legislature, 1911, Chapter 190.

91. *Smith v. Board of Examiners*, 85 N.J.L. 46; 88 A. 963; 1913; "Rules Sterilization Law Is Invalid," *New York Times*, November 19, 1913.

92. New York Laws, 1912, Chapter 445.

93. *Osborn v. Thomson*.

94. "Test Defectives' Law," *New York Times*, June 21, 1915.

95. *Osborn v. Thomson*.

96. Ibid.

97. *Smith v. Board of Examiners of Feeble-Minded*, 85 N.J. L. 46; 88 A. 963; 1913.

98. Robitscher, *Eugenic Sterilization*, 118–119.

99. Laws of New York, 1920, Chapter 619.

100. Michigan Public Acts, 1913, No. 34.

101. Myerson et al., *Eugenical Sterilization*, 13; Michigan Public Acts, 1923, No. 285; Burke Shartle, "Sterilization of Mental Defectives," *Journal of the American Institute of Criminal Law and Criminology* 16 (1926): 537–554.

102. *Smith v. Wayne Probate Judge*, 231 Mich. 409; 204 N.W. 140; 1925. The following year the same conclusions were upheld in *In re Salloum*, 236 Mich. 478; 210 N.W. 498; 1926.

103. Michigan Acts, 1929, No. 281.

104. Michigan Acts, 1943, No. 235.

105. On eugenics and coerced sterilization in Michigan, see Jeffrey Alan Hodges, "Dealing with Degeneracy: Michigan Eugenics in Context" (Ph.D. diss., Michigan State University, 2001).

106. *In re Opinion of the Justices*, 230 Ala. 543; 162 So. 123; 1935.

107. *Wyatt v. Aderholt*, 368 F. Supp. 1383; 1974.

108. Oklahoma Session Laws, 1931, Chapter 26.

109. *In re Main*, 1162 Okla. 65; 19 P.2d 153; 1933.

110. Oklahoma Session Laws 1933, Chapter 46; Oklahoma Session Laws 1935, Chapter 26.

111. *Skinner v. Oklahoma*, 115 P.2d 123; 1941.

112. *Skinner v. Oklahoma*, 316 U.S. 535; 1942.

113. Robitscher, *Eugenic Sterilization*, 118–119.

114. Idaho Session Laws, 1919, Chapter 150.

115. Laughlin, *Eugenical Sterilization*, 50.

116. Idaho Session Laws, 1925, Chapter 194; Idaho Session Laws, 1929, Chapters 68 and 285.

117. *Eugenical News* 16 (1931): 133; W. S. Dolan, "Constitutional Law—Insane and Defective Persons—Sterilization of Defectives," *Idaho Law Journal* 2 (1932): 144–145.

118. *State v. Troutman*, 50 Idaho 673; 299 P. 668; 1931.

119. Robitscher, *Eugenic Sterilization*, 118–119.

120. Idaho Session Laws, 1972, Chapter 21.

121. Idaho Statutes, Title 39, Chapter 39.

122. *Nebraska v. Gloria Cavitt et al.*, 182 Neb. 712; 157 N.W. 2d 171; 1968.

123. Robitscher, *Eugenic Sterilization*, 127.

124. *Cook v. State*, 9 Ore. App. 224, 495 P.2d 768; 1972.

125. Ibid.

126. *In re Joseph Lee Moore*, 289 N.C. 95; 221 S.E. 2d 307; 1976. See Frank Weathered, "Constitutional Law—Due Process—North Carolina Compulsory Sterilization Statute Held Constitutional against Challenge That It Constituted an Unlawful Invasion of Privacy," *Texas Tech Law Review* 8 (1976–1977): 436–445.

127. North Carolina General Statutes, 1976, Chapters 35–39(3). See also Moya Woodside, *Sterilization in North Carolina: A Sociological and Psychological Study* (Chapel Hill: University of North Carolina Press, 1950).

128. North Carolina General Statute, 1987, Chapter 35.

129. John Railey and Kevin Begos, "Board Did Its Duty, Quietly: Members from Five Governmental Areas Heard Cases Summaries and Usually Stamped Approval," http://againsttheirwill.journalnow.com. The history and impact of the state's compulsory sterilization program have been thoroughly documented in a special report from the *Winston-Salem Journal*, againsttheirwill.journalnow.com.

130. Eugenics Record, 1918–1945, 5/4/04/07, Oregon State Archives, Salem.

131. Reilly, *Surgical Solution*, 29.

132. Largent, "Greatest Curse," 205.

CHAPTER 4 — *BUCK V. BELL* AND THE FIRST ORGANIZED RESISTANCE TO COERCED STERILIZATION

1. Robert D. Johnston, "The Myth of the Harmonious City: Will Daly, Lora Little, and the Hidden Face of Progressive-Era Portland," *Oregon Historical Quarterly* 99 (1998): 275–276. See also Robert D. Johnston, *The Radical Middle Class: Populist Democracy and the Question of Capitalism in Progressive Era Portland, Oregon* (Princeton, N.J.: Princeton University Press, 2003); and Robert D. Johnston, "Contemporary Anti-Vaccination Movements in Historical Perspective," in *The Politics of Healing: Essays in Twentieth-Century History of North American Alternative Medicine* (New York: Routledge, 2004), 259-286.

2. *Buck v. Bell*, 274 U.S. 200; 47 S. Ct. 584; 1927.

3. J. W. Lockhardt, "Should Criminals Be Castrated?" *St. Louis Courier of Medicine* (1895): 136–137.

4. Corr, "Emasculation and Ovariotomy," 714–716.

5. Francis Barnes, "Vasectomy," *New England Medical Monthly* 29 (1910): 454–458.

6. Abraham Myerson, "Inheritance of Mental Diseases," *Archives of Neurology and Psychiatry* 12 (1924): 332–336.

7. "Discussion of Myerson's Inheritance of Mental Diseases," *Archives of Neurology and Psychiatry* 12 (1924): 338–339.

8. Myerson, *Inheritance of Mental Diseases*, 64–68.

9. David Heron, *A Second Study of Extreme Alcoholism in Adults with Special Reference to the Home-Office Inebriate Reformatory Data* (London: Eugenics Laboratory Publications, 1912).

10. Harry H. Laughlin, *The Legal Status of Eugenical Sterilization: History and Analysis of Litigation under the Virginia Sterilization Statute, Which Led to a Decision of the Supreme Court of the United States Upholding The Statute* (Chicago: Fred J. Ringley, 1930).

11. Abraham Myerson, "A Critique of Proposed 'Ideal' Sterilization Legislation," *Archives of Neurology and Psychiatry* 33 (1935): 453–466.

12. Michael J. Klarman, "How *Brown* Changed Race Relations: The Backlash Thesis," *Journal of American History* 81 (1994): 81–82. See also Michael J. Klarman, "*Brown*, Originalism, and Constitutional Theory: A Response to Professor McConnell," *Virginia Law Review* 81 (1995): 1881–1936; and Michael J. Klarman, *From Jim Crow to Civil Rights: The Supreme Court and the Struggle for Racial Equality* (New York: Oxford University Press, 2004).

13. Lynn, *Eugenics*, 232–233.

14. Albert W. Alschuler, *Law without Values: The Life, Work, and Legacy of Justice Holmes* (Chicago: University of Chicago Press, 1997), 65–66.

15. *In the Supreme Court of the United States*, October Term, 1926, No. 292, *Carrie Buck v. J. H. Bell.*

16. Liva Baker, *The Justice from Beacon Hill: The Life and Times of Oliver Wendell Holmes* (New York: Harper Collins, 1991), 603.

17. *Jacobsen v. Massachusetts*, 197 U.S. 11; 1905; *Buck v. Bell*, 274 U.S. 200; 47 S. Ct. 584; 1927.

18. Paul A. Lombardo, "Facing Carrie Buck," *Hasting Center Report* (March–April 2003): 14–17.

19. J. E. Coogan, "Eugenic Sterilization Hold Jubilee," *Catholic World* 177 (1953): 44.

20. Phillip Thompson, "Silent Protest: A Catholic Justice Dissents in *Buck v. Bell*," *Catholic Lawyer* 43 (2004): 125–148; *Cochran v. Louisiana Board of Education*, 281 U.S. 370; 1930 at 371.

21. *Pierce v. Society of the Sisters of the Holy Names of Jesus*, 268 U.S. 510, 1925; *Cochran v. Louisiana State Board of Education*, 281 U.S. 370, 1930; Thomas Shelley, "The Oregon School Case and the National Catholic Welfare Conference," *Catholic Historical Review* 75 (1989): 439–457.

22. David J. Danelski, *A Supreme Court Justice Is Appointed* (New York: Random House, 1964), 189.

23. Thompson, "Silent Protest," 138.

24. U.S. Supreme Court, *Proceedings of the Bar and Officers of the Supreme Court of the United States in Memory of Pierce Butler, January 27, 1940* (Washington, D.C., 1940), 14, 56.

25. William Howard Taft to Oliver Wendell Holmes Jr., April 23, 1927, Oliver Wendell Holmes Jr. Papers, Harvard Law School Library, cited in Sheldon M. Novick, *The Life of Oliver Wendell Holmes* (Boston: Little, Brown, 1989), 352–353, 477.

26. Stephen M. Donovan, "Circa Liceitatem Cujusdam Operationis Chirurgicae Proponuntur Dubia Nonnulla," *American Ecclesiastical Review* 42 (1910): 271–275; De Becker, "The Casus 'De Liceitate Vasectomiae,'" *American Ecclesiastical Review* 42 (1910): 474–475; Theodore Labouré, "De Liceitate Vasectomiae Ad Prolis Defectivae Generationem Impediendam Patratae," *American Ecclesiastical Review* 43 (1910): 70–84; Theodore Labouré, "De Vasectomia," *American Ecclesiastical Review* 43 (1910): 320–329; Theodore Labouré, "De Aliquibus Vasectomiae Liceitatem Consequentibus," *American Ecclesiastical Review* 43 (1910): 553–558; Theodore Labouré, "The Morality and Lawfulness of Vasectomy," *American Ecclesiastical Review* 44 (1911): 562–583; Austin O'Malley, "Vasectomy in Defectives," *American Ecclesiastical Review* 44 (1911): 684–705; Albert Schmitt, "Vasectomia Illicita," *American Ecclesiastical Review* 44 (1911): 679–684; Theodore Labouré, "Quid Ex Disccusione De Vasectomia Instituta Resultet," *American Ecclesiastical Review* 45 (1911): 85–98; Albert Schmitt, "The Gravity of the Mutilation Involved in Vasectomy," *American Ecclesiastical Review* 45 (1911): 360–362; Theodore Labouré, "One Last Remark on Vasectomy," *American Ecclesiastical Review* 45 (1911): 355–359; Theodore Labouré, "The Morality

and Lawfulness of Vasectomy: Is the Operation a Grave Mutilation—and What If It Is Not?" *American Ecclesiastical Review* 45 (1911): 71–77; Joseph Husslein, "The Invasion of Race Suicide and Socialism into Our Fold," *American Ecclesiastical Review* 45 (1911): 276–290; Stephen M. Donovan, "Summing Up the Discussion on Vasectomy," *American Ecclesiastical Review* 45 (1911): 313–319; Philokanon, "The Question of Vasotomy," *American Ecclesiastical Review* 45 (1911): 599–601; Austin O'Malley, "A Query about the Operation of Vasectomy," *American Ecclesiastical Review* 45 (1911): 717–721; Austin O'Malley, "De Vasectomia Duplici," *American Ecclesiastical Review* 46 (1912): 207–225.

27. De Becker, "Casus," 474.

28. "Morality and Lawfulness of Vasectomy," 562.

29. Ibid., 564.

30. Ibid., 565.

31. O'Malley, "Vasectomy in Defectives," 705.

32. Thomas Gerrard, *The Church and Eugenics* (London: S. King and Son, 1912), 9.

33. Ibid., 20; Caleb Saleeby, *Parenthood and Race Culture: An Outline of Eugenics* (New York: Moffat, Yard, 1909).

34. "Eugenics," in *The Catholic Encyclopedia: An International Work of Reference on the Constitution, Doctrine, Discipline and History of the Catholic Church* (New York: Appleton, 1907–1913), 5: 239–242.

35. John A. Ryan, *Problems of Mental Deficiency No. 3: Moral Aspects of Sterilization* (Washington, D.C.: National Catholic Welfare Conference, 1930), 7.

36. Sharon M. Leon, "'Hopelessly Entangled in Nordic Pre-supposition': Catholic Participation in the American Eugenics Society in the 1920's," *Journal of the History of Medicine and Allied Sciences* 50 (2004): 3. For an analysis of Catholic interpretations of eugenic rhetoric, see Hasian, *Rhetoric of Eugenics*, 89–111.

37. "Unjustified Sterilization," *America* (May 14, 1927): 102.

38. H. H. McClelland, "The Sterilization Fallacy," *National Catholic Welfare Conference Bulletin* (September 1927): 19–20.

39. A. R. Vonderahe, "The Sterilization of the Feebleminded," *National Catholic Welfare Conference Bulletin* (August 1927): 29–30

40. McClelland, "Sterilization Fallacy," 19.

41. "Denounce Sterilization as Outgrowth of Materialism," *National Catholic Welfare Conference Bulletin* (December 1927): 27.

42. Charles Bruehl, *Birth-Control and Eugenics in the Light of Fundamental Ethical Principles* (New York: Joseph F. Wagner, 1928), 28, 142

43. John A. Ryan, *A Living Wage: Its Ethical and Economic Aspects* (New York: Macmillan, 1906).

44. Neil Betten, "Social Catholicism and the Emergence of Catholic Radicalism in America," *Journal of Human Relations* 18 (1970): 710–727.

45. See Joseph M. McShane, *"Sufficiently Radical": Catholicism, Progressivism, and the Bishops' Program of 1919* (Washington, D.C.: Catholic University of America Press, 1986), 13.

46. John Higham, *Strangers in the Land: Patterns of American Nativism, 1860–1925* (New Brunswick, N.J.: Rutgers University Press, 1994), 39.

47. McShane, *"Sufficiently Radical,"* 17; Lawrence J. McCaffrey, *The Irish Diaspora in America* (Bloomington: Indiana University Press, 1976), 6; Thomas N. Brown, *Irish-American Nationalism, 1870–1890* (Philadelphia: J. B. Lippincott, 1966), 140.

48. McShane, *"Sufficiently Radical,"* 27.

49. Ryan, *Moral Aspects of Sterilization.*

50. Ulrich A. Hauber, *Problems of Mental Deficiency, No. 1: Inheritance of Mental Defect* (Washington, D.C.: National Catholic Welfare Conference, 1930); Charles Bernstein, *Problems of Mental Deficiency, No. 2: Social Care of the Mentally Deficient* (Washington, D.C.: National Catholic Welfare Conference, 1930); William F. Montavon, *Problems of Mental Deficiency, No. 4: Eugenic Sterilization in the Laws of the States* (Washington, D.C.: National Catholic Welfare Conference, 1930).

51. Sharon M. Leon, "'A Human Being, and Not a Mere Social Factor': Catholic Strategies for Dealing with Sterilization Statutes in the 1920's," *Church History* 73 (2004): 409.

52. Pope Pius XI, *Casti Connubii, Encyclical Letter on Christian Marriage in View of the Present Conditions, Needs, Errors and Vices that Effect the Family and Society* (1930), in *Five Great Encyclicals* (New York: Paulist Press, 1939).

53. Gene Burns, *The Moral Veto: Framing Contraception, Abortion, and Cultural Pluralism in the United States* (Cambridge: Cambridge University Press, 2005), 120. Burns cites John T. Noonan, *Contraception: A History of Its Treatment by Catholic Theologians and Canonists* (Cambridge, Mass.: Harvard University Press, 1966), 414–426, 438–447.

54. Pius XI, *Casti Connubii*, 92.

55. Ibid., 96–97.

56. Francis L. Broderick, *Right Reverend New Dealer: John A. Ryan* (New York: Macmillan, 1963), 149.

57. Patrick J. Ward, "The Grave Issue of Sterilization," *Catholic Action* 14 (1934): 7–8, 31.

58. Schmiedeler, *Sterilization*.

59. Joseph B. Lehane, *The Morality of American Civil Legislation Concerning Eugenical Sterilization* (Washington, D.C.: Catholic University of America Press, 1944).

60. Lehane, *Morality of American Civil Legislation*, 63, 80, 102–103.

61. Coogan, "Eugenic Sterilization," 47.

62. Ibid., 49.

63. Joseph D. Hassett, "Freedom and Order before God: A Catholic View," *New York University Law Review* 31 (1956): 1170–1184.

64. Paulinus F. Forsthoefel, *Religious Faith Meets Modern Science* (New York: Alba House, 1994).

65. Paulinus F. Forsthoefel, "Eugenics," in *New Catholic Encyclopedia* (Washington, D.C.: Catholic University of America, 1967), 5: 627–629.

66. On American anti-Catholicism, see Jeffrey Burton Russell, *Inventing the Flat Earth: Columbus and Modern Historians* (New York: Praeger, 1991); Andrew Greeley, *An Ugly Little Secret: Anti-Catholicism in North America* (Kansas City: Sheed Andrews and McMeel, 1977); Michael Schwartz, *The Persistent Prejudice: Anti-Catholicism in America* (Huntington, Ind.: Our Sunday Visitor, 1984); Jenny Franchot, *Roads to Rome: The Antebellum Protestant Encounter with Catholicism* (Berkeley: University of California Press, 1953); and Philip Jenkins, *The New Anti-Catholicism: The Last Acceptable Prejudice* (Oxford: Oxford University Press, 2003).

67. Frances Oswald, "Eugenical Sterilization in the United States," *American Journal of Sociology* 36 (1930): 65–73.

68. Paul Blanshard, *American Freedom and Catholic Power* (Boston: Beacon Press, 1949), 3–5.

69. Blanshard, *American Freedom*, 147–148, 154.

70. John J. Kane, *Catholic-Protestant Conflicts in America* (Chicago: Regnery, 1955), 147–154.

CHAPTER 5 — THE PROFESSIONS RETREAT

1. Myerson et al., *Eugenical Sterilization*; O'Hara and Sanks, "Eugenic Sterilization," 20–44; Legal and Socio-Economic Division of the American Medical Association, *A Reappraisal of Eugenic Sterilization Laws*, May 1960.
2. Black, *War against the Weak*.
3. C. R. H., "The Prevention of Crime, Not Merely Its Punishment," *Journal of the American Institute of Criminal Law and Criminology* 1 (1910): 9–10.
4. Giulio Battaglini, "Eugenics and Criminal Law," *Journal of the American Institute of Criminal Law and Criminology* 5 (1914): 12–15.
5. Robert H. Gault, "Heredity as a Factor in Producing the Criminal," *Journal of the American Institute of Criminal Law and Criminology* 4 (1913): 321–322.
6. Gray, "On Legislation for Sterilization," 591–592.
7. Hunter, "Sterilization of Criminals."
8. Frederick Fenning, "Sterilization Laws from a Legal Standpoint," *Journal of the American Institute of Criminal Law and Criminology* 4 (1914): 808–814.
9. H. C. Stevens, "Eugenics and Feeblemindedness," *Journal of the American Institute of Criminal Law and Criminology* 6 (1915): 190–197.
10. Edith R. Spaulding and William Healy, "Inheritance as a Factor in Criminality: A Study of a Thousand Cases of Young Repeated Offenders," *Journal of the American Institute of Criminal Law and Criminology* 4 (1914): 837–858.
11. G. E. K., "Sterilization of Criminals or Defectives," *Michigan Law Review* 12 (1914): 402.
12. Charles A. Boston, "A Protest against Laws Authorizing the Sterilization of Criminals and Imbeciles," *Journal of the American Institute of Criminal Law and Criminology* 7 (1916): 326–358.
13. Lester F. Ward, "Eugenics, Euthenics, and Eudemics," *American Journal of Sociology* 18 (1913): 737–754.
14. E.L., "Some Fallacies of Eugenics," *Journal of the American Institute of Criminal Law and Criminology* 4 (1914): 904–905.
15. F. Emory Lyon, "Race Betterment and Crime Doctors," *Journal of the American Institute of Criminal Law and Criminology* 5 (1915): 887–891.
16. L. Pierce Clark, "A Critique of the Legal, Economic and Social Status of the Epileptic," *Journal of the American Institute of Criminal Law and Criminology* 17 (1926): 218–233.
17. William A. White, "Sterilization of Criminals," *Journal of the American Institute of Criminal Law and Criminology* 8 (1917): 499–501.
18. Myerson et al., *Eugenical Sterilization*, 3.
19. Ibid., 4–5.
20. Ibid., 26.
21. Ibid., 39.
22. Ibid., 110.
23. O'Hara and Sanks, "Eugenic Sterilization," 35, 44.
24. Ibid., 36.
25. Pope Pius XI, *Casti Connubii*, 96–97.
26. O'Hara and Sanks, "Eugenic Sterilization," 38.
27. Ibid., 36–37.
28. Peter Novick, *The Holocaust in American Life* (Boston: Houghton Mifflin, 1999); Kirsten Fermaglich, *American Dreams and Nazi Nightmares: Early Holocaust Consciousness and Liberal America, 1957–1965* (Waltham, Mass.: Brandeis University Press, 2006). In reference to the "absence of the term, 'the Holocaust,'" or the "discrete narrative of the

'war against the Jews,' before the late 1960s," Fermaglich cites Gerd Korman, "The Holocaust in American Historical Writing," *Societas* 2 (1972): 259–265; Leon Jick, "The Holocaust: Its Use and Abuse within the American Public," *Yad Vashem Studies* 14 (1981): 303–318; and Jeffrey Shandler, *While America Watches: Televising the Holocaust* (New York: Oxford University Press, 1999).

29. Albert Deutsch, *The Mentally Ill in America: A History of Their Care and Treatment from Colonial Times* (Garden City, N.Y.: Doubleday, Doran, 1937), 1.

30. For a detailed discussion of nineteenth-century physicians' notions of heredity, see Zenderland, *Measuring Minds*, 150.

31. *Legal and Socio-Economic Division of the American Medical Association*, 13–14.

32. A. W. Forbes, "Is Eugenics Dead?" *Journal of Heredity* 24 (1933): 143–144.

33. C. G. Campbell, "The Present Position of Eugenics," *Journal of Heredity* 24 (1933): 144–147.

34. William H. Tucker, *The Funding of Scientific Racism: Wickliffe Draper and the Pioneer Fund* (Urbana: University of Illinois Press, 2002), 53.

35. Thomas Kuhn, *The Structure of Scientific Revolutions*, 2nd ed. (Chicago: University of Chicago Press, 1970), 136–138.

36. Steven Selden, "Education Policy and Biological Science: Genetics, Eugenics and the College Textbook, c. 1908–1931," *Teachers College Record* 87 (1985): 35–51; Selden, *Inheriting Shame.*

37. For a complete list of the textbooks examined and a description of their positions on eugenics and coerced sterilization, see the appendix.

38. Charles Robert Plunkett, *Elements of Modern Biology* (New York: Henry Holt, 1937).

39. Michael Guyer, *Animal Biology* (New York: Harper and Brothers, 1941).

40. Murville Jennings Harbaugh and Arthur Leonard Goodrich, *Fundamentals of Biology* (New York: Blakiston, 1953).

41. Garrett Hardin, *Biology: Its Principles and Implications* (San Francisco: W. H. Freeman, 1961).

42. Robert W. Korn and Ellen J. Korn, *Contemporary Perspectives in Biology* (New York: Wiley, 1971), 417.

43. Neal D. Buffaloe and J. B. Throneberry, *Concepts of Biology: A Cultural Perspective* (Englewood Cliffs, N.J.: Prentice-Hall, 1973), 299–300.

44. John Moore and Harold Slusher, *Biology: A Search for Order in Complexity* (Grand Rapids, Mich.: Zondervan, 1970), 124.

45. See, for example, Dan Gilbert, *Evolution: The Root of All Isms*, 2nd ed. (San Diego, Calif.: Danielle, 1941), 41. Gilbert wrote, "Nazism and Communism are equally expressions of the Nietzsche-Darwin theory that 'might is right and rightly so.'" Richard Weikart, *From Darwin to Hitler: Evolutionary Ethics, Eugenics, and Racism in Germany* (New York: Palgrave Macmillan, 2004). See also Tom Derosa, *Evolution's Fatal Fruit: How Darwin's Tree of Life Brought Death to Millions* (Fort Lauderdale, Fla.: Coral Ridge Ministries, 2005), and the accompanying documentary, *The Truth about Darwinism*, which includes interviews with Ann Coulter, Richard Weikart, Jonathan Wells, Phillip Johnson, and Michael Behe. For an example of this rhetoric outside of the United States, see Harun Yahya, *The Disasters Darwinism Brought to Humanity* (Scarborough, Ont.: Al-Attique, n.d.).

46. Thomas Steyaert, *Life and Patterns of Order* (New York: McGraw-Hill, 1971), 516.

47. Gideon Nelson and Gerald Robinson, *Fundamental Concepts of Biology* (New York: Wiley, 1974), 389.

48. Gairdner Moment and Helen Habermann, *Biology: A Full Spectrum* (London: Longman, 1973), 181.

49. Neil Campbell, Lawrence Mitchell, and Jane Reece, *Biology: Concepts and Connections* (San Francisco: Benjamin/Cummings, 2000), 251; James D. Watson, "The Human Genome Project: Past, Present, and Future," *Science* 248 (1990): 44–46.

50. Linus Pauling Biography, http://lpi.oregonstate.edu/lpbio/lpbio2.html.

51. Linus Pauling, "Reflections on the New Biology," *UCLA Law Review* 15 (1968): 267–272.

52. Jacob Henry Landman, *Human Sterilization: The History of the Sexual Sterilization Movement* (New York: Macmillan, 1932).

53. Jacob Henry Landman, "The History of Human Sterilization in the United States—Theory, Statute, Adjudication," *Illinois Law Review* 23 (1929): 463–480. Also published in *U.S. Law Review* 63 (1929): 48–71.

54. "Would Limit Voting to Selected Class," *New York Times*, June 7, 1931.

55. Landman, "History of Human Sterilization," 469–470.

56. "Would Limit Voting to Selected Class," 21.

57. Landman, "History of Human Sterilization," 465, 479–480.

58. *Davis v. Walton*, 74 Utah 80; 276 P. 921; 1929.

59. Landman, *Human Sterilization*, ix.

60. Ernest R. Groves, "Eugenics," *Social Forces* 13 (May 1935): 607–608; C. J. Warden, "Review of Landman's Human Sterilization," *American Journal of Psychology* 46 (October 1934): 684.

61. Sidney S. Grant, "Review of Landman's Human Sterilization," *Boston University Law Review* 12 (1932): 749–750; George Roche, "Review of Landman's Human Sterilization," *California Law Review* 22 (1933–1934): 129–131; E. S. Gosney, "Review of Landman's Human Sterilization," *Cornell Law Quarterly* 18 (1932–1933): 456–458.

62. "Review of Human Sterilization by J. H. Landman," *Idaho Law Journal* 3 (1933): 103–105.

63. Norman Himes, "Review of Landman's Human Sterilization," *Annals of the American Academy of Political and Social Science* 163 (1932): 244.

64. Himes, *History of Contraception*, 300.

65. Robert Dickinson, "Sterilizing Without Unsexing: I. Surgical Review, with Especial Reference to 5,820 Operations on Insane and Feebleminded in California," *Journal of the American Medical Association* 92 (1929): 373–379. For a critical discussion of the relationship between Dickinson's advocacy of birth control and of eugenics, see Kline, *Building a Better Race*.

66. Edward Byron Reuter, "Review of Landman's Human Sterilization," *American Journal of Sociology* 38 (1932): 286–288.

67. Edward Byron Reuter, *Handbook of Sociology* (New York: Dryden Press, 1941), 190–192. See also Edward Byron Reuter and C. W. Hart, *Introduction to Sociology* (New York: McGraw-Hill, 1933), 54, 56.

68. S. J. Holmes, "Review of Landman's Human Sterilization," *Journal of Criminal Law and Criminology* 23 (1932): 520–522.

69. Jacob Henry Landman, "Review of C. Blacker's Voluntary Sterilization," *Journal of Criminal Law and Criminology* 26 (1935): 162–162.

70. Jacob Henry Landman, "Review of Leon Whitney's The Case for Sterilization," *Journal of Criminal Law and Criminology* 26 (1935): 977–978.

71. Jacob Henry Landman, "Review of J. B. S. Haldane's Heredity and Politics," *Annals of the American Academy* 199 (1938): 276.

72. See, for example, Jacob Henry Landman, *Since 1914* (New York: Barnes and Noble, 1934).

73. Reilly, *Surgical Solution*, 30–31.

74. Bruinius, *Better for All the World*; Black, *War against the Weak*.

75. Stern, *Eugenic Nation*; Gallagher, *Breeding Better Vermonters*; Kline, *Building a Better Race*; Cogdale, *Eugenic Design*; Ordover, *American Eugenics*; Carlson, *Unfit*.

76. Lynn, *Eugenics*, vii.

CONCLUSION — THE NEW COERCED STERILIZATION MOVEMENT

1. Laura Briggs, *Reproducing Empire: Race, Sex, Science, and U.S. Imperialism in Puerto Rico* (Berkeley: University of California Press, 2002); Betsy Hartmann, *Reproductive Rights and Wrongs: The Global Politics of Population Control and Contraceptive Choice* (New York: Harper and Row, 1987).

2. Johanna Schoen, "Between Choice and Coercion: Women and the Politics of Sterilization in North Carolina," *Journal of Women's History* 13 (2001): 132–156; Schoen, *Choice and Coercion*.

3. Jane Lawrence, "The Indian Health Service and the Sterilization of Native American Women," *American Indian Quarterly* 24 (2000): 400.

4. *Genocide in Mississippi* (Atlanta: Student Nonviolent Coordinating Committee, n.d.), 3–4, 9–12.

5. Dorothy Roberts, *Killing the Black Body: Race Reproduction and the Meaning of Liberty* (New York: Pantheon Books, 1997). See also Robert G. Weisbord, *Genocide: Birth Control and the Black American* (Westport, Conn., and New York: Greenwood Press and Two Continents Publishing Group, 1975).

6. Peter Novick explored the delayed emergence of American understanding of the Holocaust in Novick, *Holocaust in American Life*.

7. Haller, *Eugenics*, 7. The first quote in this sentence is taken from the back of Rutgers's 1984 reissue in paperback of Haller's book.

8. Edwin B. Steen, *Dictionary of Biology* (New York: Barnes and Noble Books, 1971), 220.

9. Helena Curtis, *Biology*, 2nd ed. (New York: Worth, 1975), 480.

10. Diane Paul and Hamish Spencer, "Did Eugenics Rest on an Elementary Mistake?" in *The Politics of Heredity: Essays on Eugenics, Biomedicine, and the Nature-Nurture Debate* (Albany: State University of New York Press, 1998), 117–132.

11. This argument is very similar to the one offered by Paul and Spencer, "Did Eugenics Rest on an Elementary Mistake?" 129.

12. *Montana Code Annual*, §§ 53–23–101-53–23–105 (1969).

13. "Montana Eugenic Sterilization Law Applies to Retardates," *Pediatric News* (June 1974): 6–7.

14. Montana Laws, 1981, Chapter 286.

15. The number 320 was determined by adding the 256 sterilizations reported in Robitscher, *Eugenic Sterilization*, 118–119, between 1928 and 1954 with the 64 sterilizations reported between 1969 and 1974 in "Montana Eugenic Sterilization Law Applies to Retardates," 6–7.

16. On the role of Norplant in the welfare debate, see Roberts, *Killing the Black Body*, 104–149.

17. "Castration Is Proposed as Sentencing Option," *New York Times*, February 14, 1990; Tamar Lewin, "Texas Court Agrees to Castration for Rapist of 12-Year-Old Girl," *New York Times*, March 7, 1992.

18. California Penal Code, Chapter 645; Drummond Ayres Jr., "California Child Molesters Face 'Chemical Castration,'" *New York Times*, August 27, 1996; Mark Neach, "California Is on the 'Cutting Edge': Hormonal Therapy (a.k.a. 'Chemical Castration') Is Mandated for Two-Time Child Molesters," *Thomas M. Cooley Law Review* 14 (1997): 351–373; Linda Beckman, "Chemical Castration: Constitutional Issues of Due Process, Equal Protection, and Cruel and Unusual Punishment," *West Virginia Law Review* 100 (1997–1998): 853–896; Kay-Frances Brody, "A Constitutional Analysis of California's Chemical Castration Statute," *Temple Political and Civil Rights Law Review* 7 (1997–1998): 141–165; Jennifer M. Bund, "Did You Say Chemical Castration?" *University of Pittsburgh Law Review* 59 (1997–1998): 157–192; Carol Gilchrist, "An Examination of the Effectiveness of California's Chemical Castration Bill in Preventing Sex Offenders from Reoffending," *Southern California Interdisciplinary Law Journal* 7 (1998–1999): 181–203; Peter J. Gimino III, "Mandatory Chemical Castration for Perpetrators of Sex Offenses against Children: Following California's Lead," *Pepperdine Law Review* 25 (1997–1998): 67–105; Philip J. Henderson, "Section 645 of the California Penal Code: California's 'Chemical Castration' Law—A Panacea or Cruel and Unusual Punishment?" *University of San Francisco Law Review* 32 (1997–1998): 653–674; Lisa Keesling, "Practicing Medicine without a License: Legislative Attempts to Mandate Chemical Castration for Repeat Offenders," *John Marshall Law Review* 32 (1998–1999): 381–390; Raymond A. Lombardo, "California's Unconstitutional Punishment for Heinous Crimes: Chemical Castration of Sexual Offenders," *Fordham Law Review* 65 (1996–1997): 2611–2646; Karen J. Rebish, "Nipping the Problem in the Bud: The Constitutionality of California's Castration Law," *New York Law School Journal of Human Rights* 14 (1997–1998): 507–532; Jason O. Runckel, "Abuse It and Lose It: A Look at California's Mandatory Chemical Castration Law," *Pacific Law Journal* 28 (1996–1997): 547–593; Kathryn L. Smith, "Making Pedophiles Take Their Medicine: California's Chemical Castration Law," *Buffalo Public Interest Law Journal* 17 (1998–1999): 123–175; Avital Stadler, "California Injects New Life into an Old Idea: Taking a Shot at Recidivism, Chemical Castration, and the Constitution," *Emory Law Journal* 46 (1997): 1285–1326; G. L. Stelzer, "Chemical Castration and the Right to Generate Ideas: Does the First Amendment Project the Fantasies of Convicted Pedophiles?" *Minnesota Law Review* 81 (1996–1997): 1675–1709; Kris Druhm, "A Welcome Return to Draconia: California Penal Law § 645 the Castration of Sex Offenders and the Constitution," *Albany Law Review* 61 (1997–1998): 285–343.

19. Larry Helm Spaulding, "Florida's Chemical Castration Law: A Return to the Dark Ages," *Florida State University Law Review* 25 (1997–1998): 117–139; Kevin Giordano, "The Chemical Knife," archive.salon.com/health/feature/2000/03/01/castration; Caroline Wong, "Chemical Castration: Oregon's Innovative Approach to Sex Offender Rehabilitation, or Unconstitutional Punishment?" *Oregon Law Review* 80 (2001): 1–27.

20. Montana Law to Allow Injections for Rapists," *New York Times*, April 27, 1997.

21. Everett R. Holles, "Plan for Castration of 2 Child Molesters Is Opposed by Coast Medical Societies," *New York Times*, May 6, 1975; "2 Sex Offenders Sent to Prison After Plea for Castration Fails," *New York Times*, October 2, 1975.

22. *State v. Brown*, 284 S.C. 407, 326 S.E. 2d 410; 1985.

23. William E. Schmidt, "Rape Sentence: Castration or 30 Years," *New York Times*, November 26, 1983; "The Castration Option," *New York Times*, November 29, 1983; "Rapist Is Having Second Thoughts on Choosing Castration," *New York Times*, December 11, 1983; "Carolina Court Rejects Sentence of Castration," *New York Times*, October 31, 1984; "Castrated Rapist Tied to 75 New Sex Crimes," *New York Times*, November 28, 1998.

24. "Defendant Is Sterilized to Get Lesser Sentence," *New York Times*, July 22, 1986.

25. A summary description of Ashe's plea bargain can be found in "In the Courts: Georgia Woman Agrees to Sterilization to Avoid Murder Trial for Killing Infant," February 11, 2005, http://www.kaisernetwork.org/daily_reports/rep_index.cfm?hint=2&DR_ID=28111. The Web site cites Beth Warren, "Mother Chooses Sterilization over Murder Trial," *Atlanta Journal-Constitution*, February 2, 2005.

26. Lewin, "Texas Court Agrees," 1; "The Castration Option," *New York Times*, March 10, 1992; "Defendant Is Said to Change Mind on Castration," *New York Times*, March 14, 1992; Roberto Suro, "Amid Controversy, Castration Plan in Texas Rape Case Collapses," *New York Times*, March 17, 1992; "Man Who Chose Castration Gets Life Term in Sex Assault," *New York Times*, August 9, 1992.

27. Armando Villafranca and Jennifer Lenhart, "Molester Noticed by Few before Now," *Houston Chronicle*, April 8, 1996; Neach, "California Is on the 'Cutting Edge.'"

28. Verhovek, "Texas Agrees to Surgery for a Molester"; Verhovek, "Texas Frees Child Molester"; ABC News, "Texas to Monitor Child Molester with Ankle Bracelet; Lawyer Says His Client Was Castrated," May 4, 2005.

29. Council on Sex Offender Treatment, "Castration," Texas Department of State Health Services, www.dsdhs.state.tx.us/csot/csot_tcastration.shtm.

30. John Robert Hand, "Buying Fertility: The Constitutionality of Welfare Bonuses for Welfare Mothers Who Submit to Norplant Insertion," *Vanderbilt Law Review* 46 (1993): 715–754; Madeline Henley, "The Creation and Perpetuation of the Mother/Body Myth: Judicial and Legislative Enlistment of Norplant," *Buffalo Law Review* 41 (1993): 703–735; Charlotte Rutherford, "Reproductive Freedoms and African American Women," *Yale Journal of Law and Feminism* 4 (1991–1992): 255–290; Michael T. Flannery, "Norplant: The New Scarlet Letter?" *Journal of Contemporary Health Law and Policy* 8 (1992): 201–226; Dorothy E. Roberts, "Punishing Drug Addicts Who Have Babies: Women of Color, Equality, and the Right of Privacy," *Harvard Law Review* 104 (1991): 1241–1482; Janet F. Ginzberg, "Compulsory Contraception as a Condition of Probation: The Use and Abuse of Norplant," *Brooklyn Law Review* 58 (1992–1993): 979–1019; Stacey L. Arthur, "The Norplant Prescription: Birth Control Woman Control, or Crime Control?" *UCLA Law Review* 40 (1992–1993): 1–47; Kristyn M. Walker, "Judicial Control of Reproductive Freedom: The Use of Norplant as a Condition of Probation," *Iowa Law Review* 78 (1992–1993): 779–812; Melissa Burke, "The Constitutionality of the Use of the Norplant Contraceptive Device as a Condition of Probation," *Hastings Constitutional Law Quarterly* 20 (1992–1993): 207–246; William Green, "Depo-Provera, Castration, and the Probation of Rape Offenders: Statutory and Constitutional Issues," *Dayton Law Review* 12 (1986–1987): 1–26; Laurence C. Nolan, "The Unconstitutional Conditions Doctrine and Mandating Norplant for Women on Welfare Discourse," *American University Journal of Gender and Law* 15 (1994–1995): 15–37; Edward Fitzgerald, "Chemical Castration: MPS Treatment of the Sexual Offender," *American Journal of Criminal Law* 18 (1990–1991): 1–60; David S. Coale, "Norplant Bonuses and the Unconstitutional Conditions Doctrine," *Texas Law Review* 71 (1992–1993): 189–215.

31. This example is very much in line with the discussion of coercion in the conclusion of Paul's *Controlling Human Heredity*.

32. Pam Belluck, "Cash-for-Sterilization Plan Draws Addicts and Critics," *New York Times*, July 24, 1999; Jeff Stryker, "Under the Influence of Gifts, Coupons and Cash," *New York Times*, July 23, 2000; Cecilia M. Vega, "Sterilization Offer to Addicts Reopens Ethics Issue," *New York Times*, January 6, 2003.

33. "Project Prevention: Children Requiring a Caring Community," http://www.projectprevention.org.

34. Laughlin, *Eugenical Sterilization.*

35. This position is persuasively argued by Diane Paul. See Paul, *Politics of Heredity*; and Paul, *Controlling Human Heredity.*

36. Gallagher, *Breeding Better Vermonters*, 8.

Bibliography

COURT CASES

Berry et al. Constituting the Board of Parole of Iowa, et al. v. Davis, 242 U.S. 468; 37 S. Ct. 208; 61 L. Ed. 441; 1917.

Buck v. Bell, 274 U.S. 200; 47 S. Ct. 584; 1927.

Cochran v. Louisiana State Board of Education, 281 U.S. 370; 1930.

Cook v. State, 9 Ore. App. 224, 495 P.2d 768; 1972.

David v. Berry et al., 216 F. 413; 1914.

Davis v. Walton, 74 Utah 80; 276 P. 921; 1929.

Herring v. State, 119 Ga. 709, 46 S.E. 876; 1904.

In re Joseph Lee Moore, 289 N.C. 95; 221 S.E. 2d 307; 1976.

In re Main, 1162 Okla. 65; 19 P.2d 153; 1933.

In re Opinion of the Justices, 230 Ala. 543; 162 So. 123; 1935.

In re Salloum, 236 Mich. 478; 210 N.W. 498; 1926.

In re Simpson, 180 N.E. 2d 206; 1962.

In the Matter of the Sterilization of Hollis Hendrickson, 123 P.2d 322; 1942.

Jacobson v. Massachusetts, 197 U.S. 11; 1905.

Mickle v. Henrichs, Warden of Nevada State Prison, et al., 262 F. 687; 1918.

Miranda v. Arizona, 396 U.S. 868; 90 S. Ct. 140; 24 L. Ed. 2d 122; 1969.

Nebraska v. Gloria Cavitt et al., 182 Neb. 712; 157 N.W. 2d 171; 1968.

Osborn v. Thomson, Andres, and Wansboro, Composing the Examiners of Feeble-Minded Criminals and Other Defectives, 185 A.D. 902; 171 N.Y.S. 1094; 1918.

Pierce v. Society of the Sisters of the Holy Names of Jesus, 268 U.S. 510; 1925.

Skinner v. Oklahoma, 115 P.2d 123; 1941.

Skinner v. Oklahoma, 316 U.S. 535; 1942.

Smith v. Board of Examiners of Feeble-Minded, 85 N.J. L. 46; 88 A. 963; 1913.

Smith v. Wayne Probate Judge, 231 Mich. 409; 204 N.W. 140; 1925.

State v. Brown, 284 S.C. 407, 326 S.E. 2d 410; 1985.

State v. Fenner, 166 N.C. 247, 80 S.E. 970; 1914.

State v. Troutman, 50 Idaho 673; 299 P. 668; 1931.

Washington v. Feilen, 70 Wash. 65; 126 P. 75; 1912.

Williams et al. v. Smith, 190 Ind. 526; 131 N.E. 2d; 1921.

Wyatt v. Aderholt, 368 F. Supp. 1383; 1974.

ARCHIVAL SOURCES

Charles Benedict Davenport Papers, American Philosophical Society, Philadelphia.
Eugenics Record, 1918–1945, 5/4/04/07, Oregon State Archives, Salem.
Oliver Wendell Holmes Jr. Papers, Harvard Law School, Cambridge, Massachusetts.
Harry Olson Papers, Northwestern University Archives, Evanston, Illinois.
Station for Experimental Evolution Records, Carnegie Institution of Washington, Washington, D.C.

PUBLISHED SOURCES

ABC News. "Texas to Monitor Child Molester with Ankle Bracelet; Lawyer Says His Client Was Castrated." May 4, 2005.
Agar, Nicholas. *Liberal Eugenics: In Defense of Human Enhancement*. Oxford: Blackwell, 2004.
Allen, Garland. "The Eugenics Record Office at Cold Spring Harbor: An Essay on Institutional History." *Osiris* 2 (1986): 225–264.
——. "Genetics, Eugenics and Class Struggle." *Genetics* 79 (1975): 29–45.
——. "Genetics, Eugenics and Society: Internalists and Externalists in Contemporary History of Science." *Social Studies of Science* 6 (1976): 105–122.
Alschuler, Albert W. *Law without Values: The Life, Work, and Legacy of Justice Holmes*. Chicago: University of Chicago Press, 1997.
American Breeders Association: Proceedings of the Meeting Held at Columbus, Ohio, January 15–18, 1907. Washington, D.C.: American Breeders Association, 1907.
"American Work Strongly Defended: Dr. Davenport, Who Is the Centre of Criticism, Replies to Dr. Heron's Attacks: English Attack on Our Eugenics." *New York Times*, November 9, 1913.
Ames, A. C. "A Plea for Castration, as a Punishment for Crime." *Omaha Clinic* 6 (1893–1894): 343–345.
Arbogast, B. A. "Castration as a Remedy for Crime." *Denver Medical Times* 15 (August 1895): 55–58.
Arnold, Jay F. "Sociological Expediency of Sterilization Statute." *Kentucky Law Journal* 25 (1936–1937): 186–192.
——. "A Sterilization Law for Kentucky—Its Constitutionality." *Kentucky Law Journal* 24 (1935–1936): 220–229.
Arthur, Stacey L. "The Norplant Prescription: Birth Control Woman Control, or Crime Control?" *UCLA Law Review* 40 (1992–1993): 1–47.
Ayres, Drummond Ayres. "California Child Molesters Face 'Chemical Castration.'" *New York Times*, August 27, 1996.
Baker, Liva. *The Justice from Beacon Hill: The Life and Times of Oliver Wendell Holmes*. New York: Harper Collins, 1991.
Ballenger, Jesse F. *Self, Senility, and Alzheimer's Disease in Modern America: A History*. Baltimore: Johns Hopkins University Press, 2006.
Barnes, Francis. "Vasectomy." *New England Medical Monthly* 29 (1910): 454–458.
Barr, Martin W. *Mental Defectives: Their History, Treatment and Training*. Philadelphia: Blakiston, 1904.
——. "President's Annual Address." *Journal of Psycho-Asthenics* 2 (1897): 1–13.
——. "Results of Sterilization." *Journal of Psycho-Asthenics* 9 (1905): 129.
——. "Some Notes on Asexualization: With a Report of Eighteen Cases." *Journal of Nervous and Mental Diseases* 51 (1920): 231–241.

Battaglini, Giulio. "Eugenics and Criminal Law." *Journal of the American Institute of Criminal Law and Criminology* 5 (1914): 12–15.

Battey, Robert. "Castration in Mental and Nervous Diseases: A Symposium." *American Journal of Medical Sciences* 92 (1886): 483–490.

Beckman, Linda. "Chemical Castration: Constitutional Issues of Due Process, Equal Protection, and Cruel and Unusual Punishment." *West Virginia Law Review* 100 (1997–1998): 853–896.

Begos, Kevin. "Castration: Files Suggest That Punishment Was Often the Aim." http://against.theirwill.journalnow.com.

Belfield, William T. "The Sterilization of Criminals and Other Defectives by Vasectomy." *Journal of the New Mexico Medical Society* (July 1909): 21–25.

Bell, Alexander Graham. "Eugenics." *American Breeders Association Proceedings* 5 (1909): 218–220.

——. "A Few Thoughts Concerning Eugenics." *American Breeders Association Proceedings* 4 (1908): 209.

Belluck, Pam. "Cash-for-Sterilization Plan Draws Addicts and Critics." *New York Times*, July 24, 1999.

Bernard, L. L. "Review of Davenport's Naval Officers: Their Heredity and Development." *American Journal of Sociology* 25 (1919): 241–242.

Bernstein, Charles. *Problems of Mental Deficiency, No. 2: Social Care of the Mentally Deficient*. Washington, D.C.: National Catholic Welfare Conference, 1930.

Betten, Neil. "Social Catholicism and the Emergence of Catholic Radicalism in America." *Journal of Human Relations* 18 (1970): 710–727.

Black, Edwin. *War against the Weak: Eugenics and America's Campaign to Create a Master Race*. New York: Four Walls Eight Windows, 2003.

Blanshard, Paul. *American Freedom and Catholic Power*. Boston: Beacon Press, 1949.

Bogart, Henri. "Sterilization of the Unfit—The Indiana Plan." *Medical Herald* 30 (1911): 81–84.

Boston, Charles A. "A Protest against Laws Authorizing the Sterilization of Criminals and Imbeciles." *Journal of the American Institute of Criminal Law and Criminology* 7 (1916): 326–358.

Bowman, Karl M., and Bernice Engle. "The Problem of Homosexuality." *Journal of Social Hygiene* 39 (1953).

"Breeders Association Will Change Its Name." *American Breeders Magazine* 4 (1913): 177.

Briggs, Laura. *Reproducing Empire: Race, Sex, Science, and U.S. Imperialism in Puerto Rico*. Berkeley: University of California Press, 2002.

Broderick, Francis L. *Right Reverend New Dealer: John A. Ryan*. New York: Macmillan, 1963.

Brody, Kay-Frances. "A Constitutional Analysis of California's Chemical Castration Statute." *Temple Political and Civil Rights Law Review* 7 (1997–1998): 141–165.

Brown, Thomas N. *Irish-American Nationalism, 1870–1890*. Philadelphia: J. B. Lippincott, 1966.

Bruehl, Charles. *Birth-Control and Eugenics in the Light of Fundamental Ethical Principles*. New York: Joseph F. Wagner, 1928.

Bruinuis, Harry. *Better for All the World: The Secret History of Forced Sterilization and American's Quest for Racial Purity*. New York: Knopf, 2006.

Buffaloe, Neal D., and J. B. Throneberry. *Concepts of Biology: A Cultural Perspective*. Englewood Cliffs, N.J.: Prentice-Hall, 1973.

Bund, Jennifer M. "Did You Say Chemical Castration?" *University of Pittsburgh Law Review* 59 (1997–1998): 157–192.

Burke, Melissa. "The Constitutionality of the Use of the Norplant Contraceptive Device as a Condition of Probation." *Hastings Constitutional Law Quarterly* 20 (1992–1993): 207–246.

Burkhalter, Louis Wood. *Gideon Lincecum, 1793–1874: A Biography*. Austin: University of Texas Press, 1965.

Burns, Gene. *The Moral Veto: Framing Contraception, Abortion, and Cultural Pluralism in the United States*. Cambridge: Cambridge University Press, 2005.

California Department of Finance, California Statistical Abstract. www.dof.ca.gov/html/fs_data/STAT-ABS/toc.htm.

Campbell, C. G. "The Present Position of Eugenics." *Journal of Heredity* 24 (1933): 144–147.

Campbell, Neil, Lawrence Mitchell, and Jane Reece. *Biology: Concepts and Connections*. San Francisco: Benjamin/Cummings, 2000.

Carlson, Elof Axel. *The Unfit: A History of a Bad Idea*. Cold Spring Harbor, N.Y.: Cold Spring Harbor Press, 2001.

"Carolina Court Rejects Sentence of Castration." *New York Times*, October 31, 1984.

"Castrated Rapist Tied to 75 New Sex Crimes." *New York Times*, November 28, 1998.

"Castration Is Proposed as Sentencing Option." *New York Times*, February 14, 1990.

"The Castration Option." *New York Times*, March 10, 1992.

The Catholic Encyclopedia: An International Work of Reference on the Constitution, Doctrine, Discipline and History of the Catholic Church. New York: Appleton, 1907–1913.

Cave, F. C. "Report of Sterilization in the Kansas State Home for Feeble-Minded." *Journal of Psycho-Asthenics* 15 (1910–1911): 123–125.

Challener, William A. "The Law of Sexual Sterilization in Pennsylvania." *Dickinson Law Review* 57 (1952–1953): 298–306.

Chamberlin, J. E., and Sander L. Gilman, eds. *Degeneration: The Dark Side of Progress*. New York: Columbia University Press, 1985.

Clark, John E. "Asexualization of Criminals and Degenerates." *Michigan Law Journal* 6 (1897): 301–305.

Clark, L. Pierce. "A Critique of the Legal, Economic and Social Status of the Epileptic." *Journal of the American Institute of Criminal Law and Criminology* 17 (1926): 218–233.

"Climate and Eugenics." *American Breeders Magazine* 1 (1910): 183–185.

"Clippings and Comments." *Journal of Orificial Surgery* 1 (1892): 521.

Coale, David S. "Norplant Bonuses and the Unconstitutional Conditions Doctrine." *Texas Law Review* 71 (1992–1993): 189–215.

Cogdale, Christina. *Eugenic Design: Streamlining America in the 1930's*. Philadelphia: University of Pennsylvania Press, 2004.

Cole, Leon J. "The Relation of Eugenics to Euthenics." *Popular Science Monthly* 81 (1912): 475–482.

Collins, Dean. "Children of Sorrow: A History of the Mentally Retarded in Kansas." *Bulletin of the History of Medicine* 39 (1965): 59–61.

"Committee on Eugenics." *American Breeders Association Proceedings* 2 (1906): 11.

Comstock, T. Griswold. "Alice Mitchell of Memphis: 'Clippings and Comments.'" *Journal of Orificial Surgery* 1 (1892–1893): 207.

"Constitutional Law: State Sterilization Law Not Contrary to the Fourteenth Amendment Giving Due Process and Equal Protection of the Law." *Marquette Law Review* 12 (1927–1928): 73–76.

Coogan, J. E. "Eugenic Sterilization Holds Jubilee." *Catholic World* 177 (1953): 44.

Cooper, A. *Observations on the Structure and Diseases of the Testis*. London: McDowall, 1830.

Corr, A. C. "Emasculation and Ovariotomy as a Penalty for Crime and as a Reformatory Agency." *Medical Age* 13 (1895): 714–716.

Council on Sex Offender Treatment. "Castration," Texas Department of State Health Services. www.dsdhs.state.tx.us/csot/csot_tcastration.shtm.

C.R.H. "The Prevention of Crime, Not Merely Its Punishment." *Journal of the American Institute of Criminal Law and Criminology* 1 (1910): 9–10.

Curtis, Helena. *Biology*, 2nd ed. New York: Worth, 1975.

Daniel, Ferdinand Eugene. "Castration of Sexual Perverts." *Texas Medical Journal* 8 (1893): 255–271.

——. "Castration of Sexual Perverts." *Texas Medical Journal* 27 (1912): 369–385.

——. "Should Insane Criminals or Sexual Perverts Be Allowed to Procreate?" *Medico-Legal Journal* 114 (1893–1894): 275–292.

Danelski, David J. *A Supreme Court Justice Is Appointed*. New York: Random House, 1964.

Davenport, Charles. "Dr. Galloway's 'Canary Breeding.'" *Biometrika* 7 (1910): 398–400.

——. "Eugenics, a Subject for Investigation Rather Than Instruction." *American Breeders Magazine* 1 (1910): 68.

——. *Heredity in Relation to Eugenics*. New York: Henry Holt, 1911.

——. *Inheritance in Canaries*. Washington, D.C.: Carnegie Institution of Washington, 1908.

——. *Inheritance in Poultry*. Washington, D.C.: Carnegie Institution of Washington, 1906.

——. "Mendel's Law of Dichotomy in Hybrids." *Biological Bulletin* 2 (1901): 307–310.

——. "On Utilizing the Facts of Juvenile Promise and Family History in Awarding Naval Commissions to Untried Men." *Proceedings of the National Academy of Science of the United States of America* 3 (1917): 404–409.

——. "The Relation of the Association to Pure Research." *American Breeders Magazine* 1 (1910): 66–67.

——. "Report of Committee on Eugenics." *American Breeders Magazine* 1 (1910): 126–129.

——. "Zoology of the Twentieth Century." *Science* 14 (1901): 315–324.

Davenport, Charles, and Mary Theresa Scudder. *Naval Officers: Their Heredity and Development*. Washington, D.C.: Carnegie Institution of Washington, 1919.

Davies, Stanley Powell. *Social Control of the Mentally Deficient*. New York: Thomas Y. Crowell, 1930.

Dawson, B. E., ed. *Orificial Surgery: Its Philosophy, Application and Technique*. Chicago: Adams, 1912.

De Becker. "The Casus 'De Liceitate Vasectomiae.'" *American Ecclesiastical Review* 42 (1910): 474–475.

"Defendant Is Sterilized to Get Lesser Sentence." *New York Times*, July 22, 1986.

"Denounce Sterilization as Outgrowth of Materialism." *National Catholic Welfare Conference Bulletin* (December 1927): 27.

Derosa, Tom. *Evolution's Deadly Fruit: How Darwin's Tree of Life Brought Death to Millions*. Fort Lauderdale, Fla.: Coral Ridge Ministries, 2005.

Deutsch, Albert. *The Mentally Ill in America: A History of Their Care and Treatment from Colonial Times*. Garden City, N.Y.: Doubleday, Doran, 1937.

Dickinson, Robert. "Sterilizing Without Unsexing: I. Surgical Review, with Especial Reference to 5,820 Operations on Insane and Feebleminded in California." *Journal of the American Medical Association* 92 (1929): 373–379.

"Discussion of Myerson's Inheritance of Mental Diseases." *Archives of Neurology and Psychiatry* 12 (1924): 338–339.

Dolan, W. S. "Constitutional Law—Insane and Defective Persons—Sterilization of Defectives." *Idaho Law Journal* 2 (1932): 144–145.

Donald, William M. "Asexualization of Criminals and Degenerates." *Michigan Law Journal* 6 (1897): 294–297.

Donovan, Stephen M. "Circa Liceitatem Cujusdam Operationis Chirurgicae Proponuntur Dubia Nonnulla." *American Ecclesiastical Review* 42 (1910): 271–275

——. "Summing Up the Discussion on Vasectomy." *American Ecclesiastical Review* 45 (1911): 313–319.

Down, Edwin A. *The Sterilization of Degenerates*. Hartford, Conn., 1910.

"Draft of Sterilization Law for Illinois." *Journal of the American Institute of Criminal Law and Criminology* 7 (1916): 611–614.

Drake, M. J., I. W. Mills, and D. Cranston. "On the Chequered History of the Vasectomy." *British Journal of Urology* 84 (1999): 475.

Druhm, Kris. "A Welcome Return to Draconia: California Penal Law § 645 the Castration of Sex Offenders and the Constitution." *Albany Law Review* 61 (1997–1998): 285–343.

Duster, Troy. *Backdoor to Eugenics*. London: Routledge, 1990.

Eastman, Bernard Douglass. "Can Society Successfully Organize to Prevent Over-Production of Defectives and Criminals?" *Topeka Medical Journal* 9 (1897): 351–355.

Edgar, W. R. "Asexualization of Criminals and Degenerates." *Michigan Law Journal* 6 (1897): 291–294.

"Editorial: Sterilization of the Unfit." *Journal of Psycho-Asthenics* 9 (1905): 127–128.

"Editorial Comment." *Journal of the New Mexico Medical Society* (July 1909): 24–25.

E.L. "Some Fallacies of Eugenics." *Journal of the American Institute of Criminal Law and Criminology* 4 (1914): 904–905.

"Emasculation for Criminal Assaults and for Incest." *Texas Medical Journal* 22 (1907): 347.

"Emasculation of Masturbators—Is It Justifiable?" *Texas Medical Journal* 109 (1894): 243–244.

"English Expert Attacks American Eugenics Work." *New York Times*, November 9, 1913.

E.R.K. "Sterilization of Habitual Criminals and Feeble-Minded Persons." *Illinois Law Review* 5 (1910–1911): 578.

Estacio, Richard A. "Sterilization of the Mentally Disabled in Pennsylvania: Three Generations without Legislative Guidance Are Enough." *Dickinson Law Review* 92 (1987–1988): 422.

"Eugenic Sterilization in Indiana." *Indiana Law Journal* 38 (1962–1963): 275–288.

Everts, Orpheus. *Asexualization as a Penalty for Crime and Reformation of Criminals: A Paper Read before the Cincinnati Academy of Medicine, February 27, 1888*. Cincinnati: W. B. Carpenter, 1888.

Ewell, Jesse. "A Plea for Castration to Prevent Criminal Assault." *Virginia Medical Monthly* (January 11, 1907): 463–464.

Fenning, Frederick. "Sterilization Laws from a Legal Standpoint." *Journal of the American Institute of Criminal Law and Criminology* 4 (1914): 808–814.

Fermaglich, Kirsten. *American Dreams and Nazi Nightmares: Early Holocaust Consciousness and Liberal America, 1957–1965*. Waltham, Mass.: Brandeis University Press, 2006.

"The Field of Eugenics." *American Breeders Magazine* 2 (1911): 141.

Fitzgerald, Edward. "Chemical Castration: MPS Treatment of the Sexual Offender." *American Journal of Criminal Law* 18 (1990–1991): 1–60.

Flannery, Michael T. "Norplant: The New Scarlet Letter?" *Journal of Contemporary Health Law and Policy* 8 (1992): 201–226.

Forbes, A. W. "Is Eugenics Dead?" *Journal of Heredity* 24 (1933): 143–144.

Forsthoefel, Paulinus F. "Eugenics." In *New Catholic Encyclopedia*, 5:627–629. Washington, D.C.: Catholic University of America, 1967.

——. *Religious Faith Meets Modern Science*. New York: Alba House, 1994.

Funkhouser, W. D. "Eugenical Sterilization." *Kentucky Law Journal* 23 (1934–1935): 511–516.

Franchot, Jenny. *Roads to Rome: The Antebellum Protestant Encounter with Catholicism*. Berkeley: University of California Press, 1953.

Gallagher, Nancy. *Breeding Better Vermonters: The Eugenics Project in the Green Mountain State*. Hanover, N.H.: University Press of New England, 1999.

Galloway, A. Rudolf. "Canary Breeding: A Partial Analysis of Records from 1891–1909." *Biometrika* 7 (1909): 1–42.

——. "Canary Breeding: A Rejoinder to C. B. Davenport." *Biometrika* 7 (1910): 401–403.

Gault, Robert H. "Heredity as a Factor in Producing the Criminal." *Journal of the American Institute of Criminal Law and Criminology* 4 (1913): 321–322.

G.E.K. "Sterilization of Criminals or Defectives." *Michigan Law Review* 12 (1914): 402.

Gelb, Steven. "Degeneracy Theory, Eugenics, and Family Studies." *Journal of the History of the Behavioral Sciences* 26 (1990): 242–246.

Genocide in Mississippi. Atlanta: Student Nonviolent Coordinating Committee, n.d.

Gerrard, Thomas. *The Church and Eugenics*. London: S. King and Son, 1912.

Gilbert, C. B., et al. "The Prevention of Conception." *Cincinnati Medical News* 19 (1890): 303–308.

Gilbert, Dan. *Evolution: The Root of All Isms*, 2nd ed. San Diego, Calif.: Danielle, 1941.

Gilchrist, Carol. "An Examination of the Effectiveness of California's Chemical Castration Bill in Preventing Sex Offenders from Reoffending." *Southern California Interdisciplinary Law Journal* 7 (1998–1999): 181–203.

Gimino, Peter J., III. "Mandatory Chemical Castration for Perpetrators of Sex Offenses against Children: Following California's Lead." *Pepperdine Law Review* 25 (1997–1998): 67–105.

Ginzberg, Janet F. "Compulsory Contraception as a Condition of Probation: The Use and Abuse of Norplant." *Brooklyn Law Review* 58 (1992–1993): 979–1019.

Giordano, Kevin. "The Chemical Knife." archive.salon.com/health/feature/2000/03/01/castration.

Glad, John. *Future Human Evolution: Eugenics in the Twenty-first Century*. Schuylkill Haven, Pa.: Hermitage, 2006.

Goddard, Henry Herbert. *The Kallikak Family: A Study in the Heredity of Feeble-Mindedness*. New York: Macmillan, 1922.

Gordon, S. C. "Hysteria and Its Relation to Disease of the Uterine Appendages." *Journal of the American Medical Association* 6 (1886): 561–567.

Gosney, E. S. *Collected Papers on Eugenic Sterilization in California: A Critical Study of Results in 6000 Cases*. Pasadena, Calif.: Human Betterment Foundation, 1930.

——. "Review of Landman's Human Sterilization." *Cornell Law Quarterly* 18 (1932–1933): 456–458.

Gosney, E. S., and Paul Bowman Popenoe. *Sterilization for Human Betterment: A Summary of Results of 6,000 Operations in California, 1909–1929*. New York: Macmillan, 1929.

Gould, Stephen Jay. *The Mismeasure of Man*. New York: W. W. Norton, 1981.

Grant, Sidney S. "Review of Landman's Human Sterilization." *Boston University Law Review* 12 (1932): 749–750.

Gray, Albert. "Notes on the State Legislation of America in 1895." *Journal of the Society of Comparative Legislation* 1 (1896–1897): 233.

Gray, W. F. "On Legislation for Sterilization." *Journal of the American Institute of Criminal Law and Criminology* 7 (1916): 591–592.

Greeley, Andrew. *An Ugly Little Secret: Anti-Catholicism in North America.* Kansas City, Mo.: Sheed Andrews and McMeel, 1977.

Green, William. "Depo-Provera, Castration, and the Probation of Rape Offenders: Statutory and Constitutional Issues." *Dayton Law Review* 12 (1986–1987): 1–26.

Groves, Ernest R. "Eugenics." *Social Forces* 13 (May 1935): 607–608.

Gugliotta, Angela. "'Dr. Sharp with His Little Knife:' Therapeutic and Punitive Origins of Eugenic Vasectomy—Indiana, 1892–1921." *Journal of the History of Medicine* 53 (1998): 371–406.

Guyer, Michael. *Animal Biology.* New York: Harper and Brothers, 1941.

Haller, Mark. *Eugenics: Hereditarian Attitudes in American Thought.* New Brunswick, N.J.: Rutgers University Press, 1963.

Hammond, William A. "A New Substitute for Capital Punishment and Means for Preventing the Propagation of Criminals." *New York Medical Examiner* 114 (1891–1892): 190–194.

Hand, John Robert. "Buying Fertility: The Constitutionality of Welfare Bonuses for Welfare Mothers Who Submit to Norplant Insertion." *Vanderbilt Law Review* 46 (1993): 715–754.

Harbaugh, Murville Jennings, and Arthur Leonard Goodrich. *Fundamentals of Biology.* New York: Blakiston, 1953.

Hardin, Garrett. *Biology: Its Principles and Implications.* San Francisco: W. H. Freeman, 1961.

Hart, Hastings H. *The Extinction of the Defective Delinquent: A Working Program: A Paper Read before the American Prison Association at Baltimore November 12, 1912.* New York: Department of Child Helping of the Russell Sage Foundation, 1913.

——. *Sterilization as a Practical Measure: A Paper Read before the American Prison Association at Baltimore November 12th, 1912.* New York: Department of Child Helping of the Russell Sage Foundation, 1913.

——. "A Working Program for the Extinction of the Defective Delinquent." *Survey* 30 (1913): 277–279.

Hartmann, Betsy. *Reproductive Rights and Wrongs: The Global Politics of Population Control and Contraceptive Choice.* New York: Harper and Row, 1987.

Hasian, Marouf Arif. *The Rhetoric of Eugenics in Anglo-American Thought.* Athens: University of Georgia Press, 1996.

Hassencahl, Frances. "Harry H. Laughlin, 'Expert Eugenical Agent' for the House Committee on Immigration and Naturalization." Ph.D. diss., Case Western Reserve, 1970.

Hassett, Joseph D. "Freedom and Order before God: A Catholic View." *New York University Law Review* 31 (1956): 1170–1184.

Hauber, Ulrich A. *Problems of Mental Deficiency, No. 1: Inheritance of Mental Defect.* Washington, D.C.: National Catholic Welfare Conference, 1930.

Hays, Willet M. "Proposed Change in the Constitution of the American Breeders Association." *American Breeders Magazine* 1 (1910): 75.

Hegar, Alfred. "Castration in Mental and Nervous Diseases: A Symposium." *American Journal of Medical Sciences* 92 (1886): 471–483.

Henderson, Philip J. "Section 645 of the California Penal Code: California's 'Chemical Castration' Law—A Panacea or Cruel and Unusual Punishment?" *University of San Francisco Law Review* 32 (1997–1998): 653–674.

Henley, Madeline. "The Creation and Perpetuation of the Mother/Body Myth: Judicial and Legislative Enlistment of Norplant." *Buffalo Law Review* 41 (1993): 703–735.

Heron, David. "Inheritance in Canaries: A Study in Mendelism." *Biometrika* 7 (1910): 403–410.

——. *A Second Study of Extreme Alcoholism in Adults with Special Reference to the Home-Office Inebriate Reformatory Data*. London: Eugenics Laboratory Publications, 1912.

Higham, John. *Strangers in the Land: Patterns of American Nativism, 1860–1925*. New Brunswick, N.J.: Rutgers University Press, 1994.

Himes, Norman. *Medical History of Contraception*. Baltimore: Williams and Wilkins, 1936.

——. "Review of Landman's Human Sterilization." *Annals of the American Academy of Political and Social Science* (1932): 244.

Hitchcock, Charles W. "Asexualization of Criminals and Degenerates." *Michigan Law Journal* 6 (1897): 305–307.

Hodges, Jeffrey Alan. "Dealing with Degeneracy: Michigan Eugenics in Context." Ph.D. diss., Michigan State University, 2001.

Holles, Everett R. "Plan for Castration of 2 Child Molesters Is Opposed by Coast Medical Societies." *New York Times*, May 6, 1975.

Holmes, S. J. "Review of Landman's Human Sterilization." *Journal of Criminal Law and Criminology* 23 (1932): 520–522.

Houser, William H. "To Secure Immunity from Crime." *Ellingwood's Therapeutist* 2, no. 1 (May 15, 1908): 19–21.

Hughes, Charles H. "An Emasculated Homo-Sexual: His Antecedent and Post-Operative Life." *Alienist and Neurologist* 35 (1914): 277–280.

Hughes, James E. *Eugenic Sterilization in the United States: A Comparative Summary of Statutes and Review of Court Decisions*. Washington, D.C.: U.S. Public Health Service, 1940.

Hunter, Joel D. "Sterilization of Criminals (Report of Committee H of the Institute)." *Journal of the American Institute of Criminal Law and Criminology* 5 (1914): 525.

Husslein, Joseph. "The Invasion of Race Suicide and Socialism into Our Fold." *American Ecclesiastical Review* 45 (1911): 276–290.

Huston, Wendell. *Sterilization Laws: Compilation of the Sterilization Laws of Twenty-four States*. Des Moines, Iowa: Wendell Huston, 1930.

"Indiana Sterilization Bill Vetoed." *New York Times*, March 9, 1937.

Inglis, David. "Asexualization of Criminals and Degenerates." *Michigan Law Journal* 6 (1897): 298–300.

Ingram, Carl. "State Issues Apology for Policy of Sterilization." *Los Angeles Times*, March 12, 2003.

"In the Courts: Georgia Woman Agrees to Sterilization to Avoid Murder Trial for Killing Infant," February 11, 2005. http://www.kaisernetwork.org/daily_reports/rep_index. cfm?hint=2&DR_ID=28111.

In the Supreme Court of the United States, October Term, 1926, No. 292, *Carrie Buck v. J. H. Bell.*

Jackson, John P. *Science for Segregation: Race, Law and the Case against* Brown v. Board of Education. New York: New York University Press, 2005.

Jenkins, Philip. *The New Anti-Catholicism: The Last Acceptable Prejudice*. Oxford: Oxford University Press, 2003.

Jick, Leon. "The Holocaust: Its Use and Abuse within the American Public." *Yad Vashem Studies* 14 (1981): 303–318.

Johnson, David, "Out of the Ashes." *Michigan History Magazine* (January-February 2001): 22–27.

Johnson, Roswell H. "The Direct Action of the Environment." *American Breeders Association Proceedings* 5 (1909): 228.

Johnston, Robert D. "Contemporary Anti-Vaccination Movements in Historical Perspective." In *The Politics of Healing: Essays in Twentieth-Century History of North American Alternative Medicine*, 259–286. New York: Routledge, 2004.

——. "The Myth of the Harmonious City: Will Daly, Lora Little, and the Hidden Face of Progressive-Era Portland." *Oregon Historical Quarterly* 99 (1998): 275–276.

——. *The Radical Middle Class: Populist Democracy and the Question of Capitalism in Progressive Era Portland, Oregon*. Princeton, N.J.: Princeton University Press, 2003.

Jordan, David Starr. *Footnotes to Evolution*. New York: Appleton, 1898.

——. "Report on the Committee on Eugenics." *American Breeders Association Proceedings* 4 (1908): 201–208.

Kane, John J. *Catholic-Protestant Conflicts in America*. Chicago: Regnery, 1955.

Kantor, William M. "Beginnings of Sterilization in America: An Interview with Dr. Harry C. Sharp, Who Performed the First Operation Nearly Forty Years Ago." *Journal of Heredity* 28 (1937): 375.

Katz, Jonathan Ned, ed. *Gay American History: Lesbians and Gay Men in the U.S.A.*, rev. ed. New York: Penguin, 1992.

Keesling, Lisa. "Practicing Medicine without a License: Legislative Attempts to Mandate Chemical Castration for Repeat Offenders." *John Marshall Law Review* 32 (1998–1999): 381–390.

Kellar, Christian. "Asexualization—Attitude of Europeans." *Journal of Psycho-Asthenics* 9 (1905): 128–129.

Kellogg, Vernon Lyman. "Man and the Laws of Heredity." *American Breeders Association Proceedings* 5 (1909): 242.

Kennan, George. *E. H. Harriman*. Boston: Houghton-Mifflin, 1922.

Kerlin, Isaac. "President's Address." *Proceedings of the Association of Medical Officers of American Institutions for Idiotic and Feeble-Minded Persons* (1892): 274.

Kevles, Daniel. *In the Name of Eugenics: Genetics and the Uses of Human Heredity*. 1985. Reprint, Cambridge, Mass.: Harvard University Press, 1995.

Kimmelman, Barbara. "The American Breeders Association: Genetics and Eugenics in an Agricultural Context." *Social Studies of Science* 13 (1983): 163–204.

——. "A Progressive Era Discipline: Genetics at American Agricultural Colleges and Experiment Stations, 1900–1920." Ph.D. diss., University of Pennsylvania, 1987.

Kindregan, L. "Sixty Years of Compulsory Eugenic Sterilization: 'Three Generations of Imbeciles' and the Constitution of the United States." *Chicago-Kent Law Review* 123 (1966): 123–143.

Kirkley, C. A. "Gynecological Observations in the Insane." *Journal of the American Medical Association* 19 (1892): 553–555.

Klarman, Michael J. "*Brown*, Originalism, and Constitutional Theory: A Response to Professor McConnell." *Virginia Law Review* 81 (1995): 1881–1936.

——. *From Jim Crow to Civil Rights: The Supreme Court and the Struggle for Racial Equality*. New York: Oxford University Press, 2004.

——. "How *Brown* Changed Race Relations: The Backlash Thesis." *Journal of American History* 81 (1994): 81–118.

Kline, Wendy. *Building a Better Race: Gender, Sexuality, and Eugenics from the Turn of the Century to the Baby Boom*. Berkeley: University of California Press, 2001.

Kopp, Marie E. "Surgical Treatment as Sex Crime Prevention Measure." *Journal of Criminal Law and Criminology* 29 (1938): 697.

Korman, Gerd. "The Holocaust in American Historical Writing." *Societas* 2 (1972): 259–265.

Korn, Robert W., and Ellen J. Korn. *Contemporary Perspectives in Biology*. New York: Wiley, 1971.

Kuhn, Thomas. *The Structure of Scientific Revolutions*, 2nd ed. Chicago: University of Chicago Press, 1970.

Labouré, Theodore. "De Aliquibus Vasectomiae Liceitatem Consequentibus." *American Ecclesiastical Review* 43 (1910): 553–558.

——. "De Liceitate Vasectomiae Ad Prolis Defectivae Generationem Impediendam Patratae." *American Ecclesiastical Review* 43 (1910): 70–84.

——. "De Vasectomia." *American Ecclesiastical Review* 43 (1910): 320–329.

——. The Morality and Lawfulness of Vasectomy." *American Ecclesiastical Review* 44 (1911): 562–583.

——. "One Last Remark on Vasectomy." *American Ecclesiastical Review* 45 (1911): 355–359.

——. "Quid Ex Disccusione De Vasectomia Instituta Resultet." *American Ecclesiastical Review* 45 (1911): 85–98.

Landman, Jacob Henry. "The History of Human Sterilization in the United States—Theory, Statute, Adjudication." *Illinois Law Review* 23 (1929): 463–480.

——. "The History of Human Sterilization in the United States—Theory, Statute, Adjudication." *U.S. Law Review* 63 (1929): 48–71.

——. *Human Sterilization: The History of the Sexual Sterilization Movement*. New York: Macmillan, 1932.

——. "Review of C. P. Blacker's Voluntary Sterilization." *Journal of Criminal Law and Criminology* 26 (1935): 161–162.

——. "Review of J.B.S. Haldane's Heredity and Politics." *Annals of the American Academy* 199 (1938): 276.

——. "Review of Leon Whitney's The Case for Sterilization." *Journal of Criminal Law and Criminology* 26 (1935): 977–978.

——. *Since 1914*. New York: Barnes and Noble, 1934.

Largent, Mark A. "'The Greatest Curse of the Race:' Eugenic Sterilization in Oregon, 1909–1983." *Oregon Historical Quarterly* 103 (2002): 188–209.

Larson, Edward. *Sex, Race, and Science: Eugenics in the Deep South*. Baltimore: Johns Hopkins University Press, 1995.

Laughlin, Harry H. *Eugenical Sterilization in the United States*. Chicago: Psychopathic Laboratory of the Municipal Court of Chicago, 1922.

——. "The Eugenics Record Office at the End of Twenty-seven Months' Work." *Report of the Eugenics Record Office* 1 (1912).

——. "The Fundamental Biological and Mathematical Principles Underlying Chromosomal Descent and Recombination in Human Heredity." In *Research Studies of Crime as Related to Heredity*, 74–82. Chicago: Municipal Court of Chicago, 1925.

——. *The Legal Status of Eugenical Sterilization: History and Analysis of Litigation under the Virginia Sterilization Statute, Which Led to a Decision of the Supreme Court of the United States Upholding The Statute*. Chicago: Fred J. Ringley, 1930.

Lawrence, Jane. "The Indian Health Service and the Sterilization of Native American Women." *American Indian Quarterly* 24 (2000): 400–419.

Legal and Socio-Economic Division of the American Medical Association. *A Reappraisal of Eugenic Sterilization Laws*. May 1960.

"Legislation against the Propagation of Disease." *West Virginia Bar* 3 (1896): 160–161.

Lehane, Joseph B. *The Morality of American Civil Legislation Concerning Eugenical Sterilization*. Washington, D.C.: Catholic University of America Press, 1944.

Leon, Sharon M. "'Hopelessly Entangled in Nordic Pre-Supposition:' Catholic Participation in the American Eugenics Society in the 1920s." *Journal of the History of Medicine and Allied Sciences* 50 (2004): 3–49.

——. "'A Human Being, and Not a Mere Social Factor': Catholic Strategies for Dealing with Sterilization Statutes in the 1920s." *Church History* 73 (2004): 383–411.

Lewin, Tamar. "Texas Court Agrees to Castration for Rapist of 13-Year-Old Girl." *New York Times*, March 7, 1992.

Lightner, Clarence. "Asexualization of Criminals and Degenerates." *Michigan Law Journal* 6 (1897): 312–315.

Lindman, Frank T., and Donald M. McIntyre. *The Mentally Disabled and the Law.* Chicago: University of Chicago Press, 1961.

"Linus Pauling Biography." http://lpi.oregonstate.edu/lpbio/lpbio2.html.

Lockhardt, J. W. "Should Criminals Be Castrated?" *St Louis Courier of Medicine* (1895): 136–137.

Lombardo, Paul A. "Eugenic Sterilization in Virginia: Aubrey Strode and the Case of *Buck v. Bell.*" Ph.D. diss., University of Virginia, 1982.

——. "Facing Carrie Buck." *Hasting Center Report* (March–April 2003): 14–17.

——. "Involuntary Sterilization in Virginia: From *Buck v. Bell* to *Poe v. Lynchburg.*" *Developments in Mental Health Law* 3 (1983): 17–21.

——. "Three Generations, No Imbeciles: New Light on *Buck v. Bell.*" *NYU Law Review* 60 (1985): 30–62.

Lombardo, Raymond A. "California's Unconstitutional Punishment for Heinous Crimes: Chemical Castration of Sexual Offenders." *Fordham Law Review* 65 (1996–1997): 2611–2646.

Ludmerer, Kenneth. *Genetics and American Society: A Historical Appraisal.* Baltimore: Johns Hopkins University Press, 1972.

Lydston, G. Frank. *Addresses and Essays.* Louisville: Renz and Henry, 1892.

——. *The Diseases of Society: The Vice and Crime Problem.* Philadelphia: J. B. Lippincott, 1904.

——. "Sexual Perversion, Satyriasis and Nymphomania." *Medical and Surgical Reporter* 61 (1889): 253.

Lydston, G. Frank, and Hunter McGuire. *Sexual Crimes among the Southern Negroes.* Louisville: Renz and Henry, 1893.

Lynn, Richard. *Eugenics: A Reassessment.* Westport, Conn.: Praeger, 2001.

Lyon, F. Emory. "Race Betterment and Crime Doctors." *Journal of the American Institute of Criminal Law and Criminology* 5 (1915): 887–891.

MacDowell, E. Carleton. "Charles Benedict Davenport: A Study of Conflicting Influences." *Bios* 17 (1946): 3–50.

Mack, Pamela. "The Early Years of Cold Spring Harbor Station for Experimental Evolution." Unpublished manuscript, 1979. American Philosophical Society.

Manganaro, Christine. "Eugenicist as Patient Advocate: Therapeutic Sterilization in Washington State." Honors thesis, University of Puget Sound, 2003.

"Man Who Chose Castration Gets Life Term in Sex Assault." *New York Times*, August 9, 1992.

Mason, Russell Z. Mason. "The Duty of the State in Its Treatment of the Deaf and Dumb, the Blind, the Idiotic, the Crippled and Deformed, and the Insane." *Transactions of the Wisconsin Academy of Sciences, Arts and Letters* 4 (1876–1877): 27–29.

McCaffrey, Lawrence J. *The Irish Diaspora in America.* Bloomington: Indiana University Press, 1976.

McClelland, H. H. "The Sterilization Fallacy." *National Catholic Welfare Conference Bulletin* (September 1927): 19–20.

McCormick, C. O. "Is the Indiana 1935 Sterilization of the Insane Act Functioning?" *Journal of the Indiana State Medical Association* 42 (1949): 919–920.

McShane, Joseph M. *"Sufficiently Radical": Catholicism, Progressivism, and the Bishops' Program of 1919*. Washington, D.C.: Catholic University of America Press, 1986.

Mears, J. Ewing. "Asexualization as a Remedial Measure in the Relief of Certain Forms of Mental, Moral and Physical Degeneration." *Boston Medical and Surgical Journal* 161 (1909): 584–586.

Metcalf, Maynard M. "Eugenics and Euthenics." *Popular Science Monthly* 84 (1914): 383–389.

Michigan Legislature. *Journal of the House of Representatives of the State of Michigan, 1897*. Detroit: Morse and Bagg, 1897.

Millikin, Mark. "The Proposed Castration of Criminals and Sexual Perverts." *Cincinnati Lancet-Clinic* 33 (1894): 185–190.

Moment, Gairdner, and Helen Habermann. *Biology: A Full Spectrum*. London: Longman, 1973.

"Montana Eugenic Sterilization Law Applies to Retardates." *Pediatric News* (June 1974): 6–7.

"Montana Law to Allow Injections for Rapists." *New York Times*, April 27, 1997.

Montavon, William F. *Problems of Mental Deficiency, No. 4: Eugenic Sterilization in the Laws of the States*. Washington, D.C.: National Catholic Welfare Conference, 1930.

Moore, John N., and Harold Schultz Slusher. *Biology: A Search for Order in Complexity*. Grand Rapids, Mich.: Zondervan, 1970.

"The Morality and Lawfulness of Vasectomy: Is the Operation a Grave Mutilation—and What if It Is Not?" *American Ecclesiastical Review* 45 (1911): 71–77.

Morley, Charles S. "Asexualization of Criminals and Degenerates." *Michigan Law Journal* 6 (1897): 307–310.

"The Movement to Study Eugenics." *American Breeders Magazine* 1 (1910): 143.

Mowry, George E. "Review: Eugenics and the Progressives." *American Historical Review* 74 (1969): 1740–1741.

Mulheron, J. J. "Asexualization of Criminals and Degenerates." *Michigan Law Journal* 6 (1897): 301.

"Mutilation." *Winfield Daily Courier*, August 24, 1894.

M'Vey, R. E. "Crime—Its Physiology and Pathogenesis. How Can Medical Men Aid in Its Prevention?" *Kansas Medical Journal* 2 (1890): 499–504.

Myerson, Abraham. "Certain Medical and Legal Phases of Eugenic Sterilization." *Yale Law Journal* 52 (1943): 618.

———. "A Critique of Proposed 'Ideal' Sterilization Legislation." *Archives of Neurology and Psychiatry* 33 (1935): 453–466.

———. "Inheritance of Mental Diseases." *Archives of Neurology and Psychiatry* 12 (1924): 332–336.

———. *The Inheritance of Mental Diseases*. Baltimore: Williams and Williams, 1925.

Myerson, Abraham, James B. Ayer, Tracy Putnam, Clyde Keeler, and Leo Alexander. *Eugenical Sterilization: A Reorientation of the Problem*. New York: Macmillan, 1936.

Neach, Mark J. "California Is on the 'Cutting Edge': Hormonal Therapy (a.k.a. 'Chemical Castration') Is Mandated for Two-Time Child Molesters." *Thomas M. Cooley Law Review* 14 (1997): 351–373.

Nedrud, Duane R. "Sterilization—Scope of the State's Power to Use Sterilization on Mental Defectives and Criminals—Operation of North Dakota Statute." *North Dakota Bar Briefs* 26 (1950): 190.

Nelson, Gideon, and Gerald Robinson. *Fundamental Concepts of Biology*. New York: Wiley, 1974.

"The New Eugenics Section." *American Breeders Magazine* 1 (1910): 153.

"The New Magazine." *American Breeders Magazine* 1 (1910): 61.

Nolan, Laurence C. "The Unconstitutional Conditions Doctrine and Mandating Norplant for Women on Welfare Discourse." *American University Journal of Gender and Law* 15 (1994–1995): 15–37.

Noonan, John T. *Contraception: A History of Its Treatment by Catholic Theologians and Canonists*. Cambridge, Mass.: Harvard University Press, 1966.

"Notes." *Albany Law Journal* 55 (1897): 375–376.

"Notes and Legislation." *Iowa Law Review* 35 (1949–1950): 251–304.

Nordau, Max Simon. *Degeneration*. New York: D. Appleton, 1895.

Novick, Peter. *The Holocaust in American Life*. Boston: Houghton Mifflin, 1999.

Novick, Sheldon. *Honorable Justice: The Life of Oliver Wendell Holmes*. Boston: Little, Brown, 1989.

Ochsner, A. J. "Surgical Treatment of Habitual Criminals." *Journal of the American Medical Association* 32 (April 1899): 867–868.

O'Hara, James B., and T. Howland Sanks, "Eugenic Sterilization." *Georgetown Law Journal* 45 (1956): 20–44.

Olson, Harry. "President's Address to the Eleventh Annual Meeting of the Eugenics Research Association, Cold Springs Harbor, L.I., June 16, 1923." In *Research Studies of Crime as Related to Heredity*, 9–61. Chicago: Municipal Court of Chicago, 1925.

O'Malley, Austin. "De Vasectomia Duplici." *American Ecclesiastical Review* 46 (1912): 207–225.

——. "A Query about the Operation of Vasectomy." *American Ecclesiastical Review* 45 (1911): 717–721.

——. "Vasectomy in Defectives." *American Ecclesiastical Review* 44 (1911): 684–705.

Ordover, Nancy. *American Eugenics: Race, Queer Anatomy, and the Science of Nationalism*. Minneapolis: University of Minnesota Press, 2003.

Oswald, Frances. "Eugenical Sterilization in the United States." *American Journal of Sociology* 36 (1930): 65–73.

"Our Supplement." *American Lawyer* 5 (1897): 299.

Owen, Orville W. "Asexualization of Criminals and Degenerates." *Michigan Law Journal* 6 (1897): 311–312.

Owens-Adair, Bethenia. "Letter to the Editor." *Oregonian*, March 11, 1904.

Paul, Diane B. *Controlling Human Heredity: 1865 to the Present*. Atlantic Highlands, N.J.: Humanities Press, 1995.

——. *The Politics of Heredity: Essays on Eugenics, Biomedicine, and the Nature-Nurture Debate*. Albany: State University of New York Press, 1998.

Paul, Diane B., and Hamish Spencer. "Did Eugenics Rest on an Elementary Mistake?" In *The Politics of Heredity: Essays on Eugenics, Biomedicine, and the Nature-Nurture Debate*, 117-132. Albany: State University of New York Press, 1998.

Paul, Julius. "Population 'Quality' and 'Fitness for Parenthood' in the Light of State Eugenic Sterilization Experience." *Population Studies* 21 (1967): 297.

Pauling, Linus. "Reflections on the New Biology." *UCLA Law Review* 15 (1968): 267–272.

Pauly, Philip J. *Biologists and the Promise of American Life*. Princeton, N.J.: Princeton University Press, 2000.

"The Pedagogics of Eugenics." *American Breeders Magazine* 3 (1912): 224.

Pernick, Martin. *The Black Stork: Eugenics and the Death of "Defective" Babies in American Medicine Motion Pictures since 1915*. Oxford: Oxford University Press, 1996.

Philokanon. "The Question of Vasotomy." *American Ecclesiastical Review* 45 (1911): 599–601.

Pickens, Donald. *Eugenics and the Progressives*. Nashville, Tenn.: Vanderbilt University Press, 1968.

Plunkett, Charles Robert. *Elements of Modern Biology*. New York: Henry Holt, 1937.

Pope Pius XI. *Casti Connubii, Encyclical Letter on Christian Marriage in View of the Present Conditions, Needs, Errors and Vices that Effect the Family and Society*, 1930. In *Five Great Encyclicals*. New York: Paulist Press, 1939.

Porter, Theodore M. *Karl Pearson: The Scientific Life in a Statistical Age*. Princeton, N.J.: Princeton University Press, 2004.

Pratt, Edwin Hartley. "A Monthly Series of Articles upon the Orificial Philosophy Written for the Journal of Orificial Surgery." *Journal of Orificial Surgery* 1 (1892): 22–27.

——. *Orificial Surgery and Its Application to the Treatment of Chronic Diseases*. Chicago: Halsey Brothers, 1887.

"Preachers and Eugenics." *American Breeders Magazine* 4 (1913): 63.

"Proceedings of the First Meeting of the American Breeders Association Held at St. Louis, Missouri, December 29 and 30, 1903." *American Breeders Association Proceedings* 1 (1905): 9–15.

"Project Prevention: Children Requiring a Caring Community." http://www.project-prevention.org.

"Race Suicide for Social Parasites." *Journal of the American Medical Association* 50 (1908): 55–56.

Railey, John, and Kevin Begos. "Board Did Its Duty, Quietly: Members from Five Governmental Areas Heard Cases Summaries and Usually Stamped Approval." http://against-theirwill.journalnow.com.

"Rapist Is Having Second Thoughts on Choosing Castration." *New York Times*, December 11, 1983.

Rebish, Karen J. "Nipping the Problem in the Bud: The Constitutionality of California's Castration Law." *New York Law School Journal of Human Rights* 14 (1997–1998): 507–532.

Reilly, Philip. *The Surgical Solution: A History of Involuntary Sterilization in the United States*. Baltimore: Johns Hopkins University Press, 1991.

"Removal of the Ovaries as a Therapeutic Measure in Public Institutions for the Insane." *Journal of the American Medical Association* 20 (1893): 135–137.

Rentoul, Robert Reid. *Proposed Sterilization of Certain Mental and Physical Degenerates: An Appeal to Asylum Managers and Others*. London: Walter Scott, 1903.

——. "Proposed Sterilization of Certain Mental Degenerates." *American Journal of Sociology* 12 (1906): 319–327.

——. *Race Culture: Or, Race Suicide (A Plea for the Unborn)*. London: Walter Scott, 1906.

Research Studies of Crime as Related to Heredity. Chicago: Municipal Court of Chicago, 1925.

Reuter, Edward Byron. *Handbook of Sociology*. New York: Dryden Press, 1941.

——. "Review of Landman's Human Sterilization." *American Journal of Sociology* 38 (1932): 286–288.

Reuter, Edward Bryon, and C. W. Hart. *Introduction to Sociology*. New York: McGraw-Hill, 1933.

"Review of Human Sterilization by J. H. Landman." *Idaho Law Journal* 3 (1933): 103–105.

R.H.G. "State Laws Regulating Marriage of the Unfit." *Journal of the American Institute of Criminal Law and Criminology* 4 (1913): 423–425.

Riddle, Oscar. "Charles Benedict Davenport." *Biographical Memoirs of the National Academy of Sciences* 25 (1949): 75–110.

Risley, S. D. "Is Asexualization Ever Justifiable in the Case of Imbecile Children?" *Journal of Psycho-Asthenics* 9 (1905): 93–98.

Roberts, Dorothy. *Killing the Black Body: Race, Reproduction, and the Meaning of Liberty.* New York: Pantheon Books, 1997.

——. "Punishing Drug Addicts Who Have Babies: Women of Color, Equality, and the Right of Privacy." *Harvard Law Review* 104 (1991): 1241–1482.

Robitscher, Jonas. *Eugenic Sterilization.* Springfield, Ill.: Thomas, 1973.

Roby, Henry. "Family Doctor." *Kansas Farmer.* Reprinted in "Emasculation of Masturbators—Is It Justifiable?" *Texas Medical Journal* 109 (1894): 239–241.

Roche, George. "Review of Landman's Human Sterilization." *California Law Review* 22 (1933–1934): 129–131.

Rosen, Christine. *Preaching Eugenics: Religious Leaders and the American Eugenics Movement.* Oxford: Oxford University Press, 2004.

Rosenberg, Charles. *No Other Gods: On Science and American Social Thought.* Baltimore: Johns Hopkins University Press, 1975.

"Rules Sterilization Law Is Invalid." *New York Times*, November 19, 1913.

Runckel, Jason O. "Abuse It and Lose It: A Look at California's Mandatory Chemical Castration Law." *Pacific Law Journal* 28 (1996–1997): 547–593.

Russell, Jeffrey Burton. *Inventing the Flat Earth: Columbus and Modern Historians.* New York: Praeger, 1991.

Rutherford, Charlotte. "Reproductive Freedoms and African American Women." *Yale Journal of Law and Feminism* 4 (1991–1992): 255–290.

Rutkow, Ira M. "Edwin Hartley Pratt and Orificial Surgery: Unorthodox Surgical Practice in Nineteenth Century United States." *Surgery* 114 (1993): 558–563.

——. "Orificial Surgery." *Archives of Surgery* 136 (2001): 1088.

Ryan, John A. *A Living Wage: Its Ethical and Economic Aspects.* New York: Macmillan, 1906.

——. *Problems of Mental Deficiency No. 3: Moral Aspects of Sterilization.* Washington, D.C.: National Catholic Welfare Conference, 1930.

Saleeby, Caleb. *Parenthood and Race Culture: An Outline of Eugenics.* New York: Moffat, Yard, 1909.

Schmidt, William E. "Rape Sentence: Castration or 30 Years." *New York Times*, November 26, 1983.

Schmiedeler, Edgar. *Sterilization in the United States: Introduction, Legal Status, the Question of Heredity, the Moral Question, the Solution.* Washington, D.C.: Family Life Bureau, National Catholic Welfare Conference, 1943.

Schmitt, Albert. "The Gravity of the Mutilation Involved in Vasectomy." *American Ecclesiastical Review* 45 (1911): 360–352.

——. "Vasectomia Illicita." *American Ecclesiastical Review* 44 (1911): 679–684.

Schwartz, Michael. *The Persistent Prejudice: Anti-Catholicism in America.* Huntington, Ind.: Our Sunday Visitor, 1984.

Schoen, Johanna. "Between Choice and Coercion: Women and the Politics of Sterilization in North Carolina." *Journal of Women's History* 13 (2001): 132–156.

——. *Choice and Coercion: Birth Control, Sterilization, and Abortion in Public Health and Welfare.* Chapel Hill: University of North Carolina Press, 2005.

Seaton, Frederick D. "The Long Road toward 'The Right Thing to Do.'" *Kansas History: A Journal of the Central Plains* 27 (2004–2005): 250–265.

"Secretary's Report." *American Breeders Association Proceedings* 8 (1912): 324–335.

"The Section Organization." *American Breeders Magazine* 1 (1910): 143.

Selden, Steven. "Education Policy and Biological Science: Genetics, Eugenics and the College Textbook, c. 1908–1931." *Teachers College Record* 87 (1985): 35–51.

——. *Inheriting Shame: The Story of Eugenics and Racism in America*. New York: Teachers College Press, 1999.

Sellin, Thorsten. "Hastings Hornell Hart." *Journal of Criminal Law and Criminology* 23 (1932): 166.

Sengoopta, C. "'Dr. Steinach Coming to Make Old Young!': Sex Glands, Vasectomy, and the Quest for Rejuvenation in the Roaring Twenties." *Endeavor* 27 (2003): 122–126.

Seven Great Encyclicals. Glen Rock, N.J.: Paulist Press, 1963.

Shandler, Jeffrey. *While America Watches: Televising the Holocaust*. New York: Oxford University Press, 1999.

Sharp, Harry C. "The Indiana Idea of Human Sterilization." *Southern California Practitioner* 24 (1909): 549–551.

——. "The Severing of the Vasa Deferentia and Its Relation to the Neuropsychopathic Constitution." *New York Medical Journal* (1902): 413.

——. "Vasectomy as a Means of Preventing Procreation in Defectives." *Journal of the American Medical Association* 53 (1909): 1897–1902.

Shartle, Burke. "Sterilization of Mental Defectives." *Journal of the American Institute of Criminal Law and Criminology* 16 (1926): 537–554.

Shelley, Thomas. "The Oregon School Case and the National Catholic Welfare Conference." *Catholic Historical Review* 75 (1989): 439–457.

Sim, F. L. "Asexualization for the Prevention of Crime and the Arrest of the Propagation of Criminals." *Medical Society of Tennessee* (1894): 100–114.

——. "Forensic Psychiatry: Alice Mitchell Adjudged Insane." *Memphis Medical Monthly* 12 (1892): 377–428.

Smith, Jessie Spaulding. "Marriage, Sterilization and Commitment Laws Aimed at Decreasing Mental Deficiency." *Journal of the American Institute of Criminal Law and Criminology* 5 (1914): 364–370.

Smith, Kathryn L. "Making Pedophiles Take Their Medicine: California's Chemical Castration Law." *Buffalo Public Interest Law Journal* 17 (1998–1999): 123–175.

Spaulding, Edith R., and William Healy. "Inheritance as a Factor in Criminality: A Study of a Thousand Cases of Young Repeated Offenders." *Journal of the American Institute of Criminal Law and Criminology* 4 (1914): 837–858.

Spaulding, Larry Helm. "Florida's Chemical Castration Law: A Return to the Dark Ages." *Florida State University Law Review* 25 (1997–1998): 117–139.

Spence, James. "The Law of the Crime against Nature." *North Carolina Law Review* 32 (1953–1954): 312–324.

Spencer, Edward W. "Some Phases of Marriage Law and Legislation from a Sanitary and Eugenic Standpoint." *Yale Law Journal* 25 (1915): 58–73.

Spencer, Hamish G., and Diane B. Paul. "The Failure of a Scientific Critique: David Heron, Karl Pearson, and Mendelian Eugenics." *British Journal of the History of Science* 31 (1998): 441–452.

Stadler, Avital. "California Injects New Life into an Old Idea: Taking a Shot at Recidivism, Chemical Castration, and the Constitution." *Emory Law Journal* 46 (1997): 1285–1326.

Steen, Edwin B. *Dictionary of Biology*. New York: Barnes and Noble Books, 1971.

Steiner, R. L. "Eugenics in Oregon." *Northwest Medicine* 26 (December 1927): 594–597.

Stelzer, G. L. "Chemical Castration and the Right to Generate Ideas: Does the First Amendment Project the Fantasies of Convicted Pedophiles?" *Minnesota Law Review* 81 (1996–1997): 1675–1709.

"Sterilization of Criminals." *Fortune* (June 1937): 106.

Stern, Alexandra Minna. *Eugenic Nation: Faults and Frontiers of Better Breeding in Modern America*. Berkeley: University of California Press, 2005.

Stern, Curt. *Principles of Human Genetics*. San Francisco: W. H. Freeman, 1949.

Stevens, H. C. "Eugenics and Feeblemindedness." *Journal of the American Institute of Criminal Law and Criminology* 6 (1915): 190–197.

Steyaert, Thomas. *Life and Patterns of Order*. New York: McGraw-Hill, 1971.

Stryker, Jeff. "Under the Influence of Gifts, Coupons and Cash." *New York Times*, July 23, 2000.

Stuver, E. "Asexualization for the Limitation of Disease and the Prevention and Punishment of Crime." *Transactions of the Colorado State Medical Society* (1895): 327–336.

Suro, Roberto. "Amid Controversy, Castration Plan in Texas Rape Case Collapses." *New York Times*, March 17, 1992.

"Test Defectives' Law." *New York Times*, June 21, 1915.

"These Are Times of Scientific Ideals." *American Breeders Magazine* 4 (1913): 61.

Thompson, Phillip. "Silent Protest: A Catholic Justice Dissents in *Buck v. Bell*." *Catholic Lawyer* 43 (2004): 125–148.

Tucker, William H. *The Founding of Scientific Racism: Wickliffe Draper and the Pioneer Fund*. Urbana: University of Illinois Press, 2002.

"2 Sex Offenders Sent to Prison After Plea for Castration Fails." *New York Times*, October 2, 1975.

United States Supreme Court, Proceedings of the Bar and Officers of the Supreme Court of the United States in Memory of Pierce Butler, January 27, 1940. Washington, D.C., 1940.

"Unjustified Sterilization." *America* (May 14, 1927): 102.

Van Roden, Edward Leroy. "Sterilization of Abnormal Persons as Punishment for and Prevention of Crimes." *Temple Law Quarterly* 23 (1949): 99–106.

"Vasectomy for Confirmed Criminals and Defectives." *Journal of the American Medical Association* 53 (1909): 1114.

Vecoli, Rudolph J. "Sterilization: A Progressive Measure?" *Wisconsin Magazine of History* 43 (1960): 190–202.

Vega, Cecilia M. "Sterilization Offer to Addicts Reopens Ethics Issue." *New York Times*, January 6, 2003.

Verhovek, Sam Howe. "Texas Agrees to Surgery for a Molester." *New York Times*, April 5, 1996.

——. "Texas Frees Child Molester Who Warns of New Crimes." *New York Times*, April 9, 1996.

Villafrana, Armando, and Jennifer Lenhart. "Molester Noticed by Few before Now." *Houston Chronicle*, April 8, 1996.

Vonderahe, A. R. "The Sterilization of the Feebleminded." *National Catholic Welfare Conference Bulletin* (August 1927): 29–30

Walker, Kristyn M. "Judicial Control of Reproductive Freedom: The Use of Norplant as a Condition of Probation." *Iowa Law Review* 78 (1992–1993): 779–812.

Ward, Lester F. "Eugenics, Euthenics, and Eudemics." *American Journal of Sociology* 18 (1913): 737–754.

Ward, Patrick J. "The Grave Issue of Sterilization." *Catholic Action* 14 (1934): 7–8, 31.

Warden, C. J. "Review of Landman's Human Sterilization." *American Journal of Psychology* 46 (1934): 684.

Warren, Beth. "Mother Chooses Sterilization over Murder Trial." *Atlanta Journal-Constitution*, February 2, 2005.

Watson, James D. "The Human Genome Project: Past, Present, and Future." *Science* 248 (1990): 44–46.

Weathered, Frank. "Constitutional Law—Due Process—North Carolina Compulsory Sterilization Statute Held Constitutional against Challenge That It Constituted an Unlawful Invasion of Privacy." *Texas Tech Law Review* 8 (1976–1977): 436–445.

Weikart, Richard. *From Darwin to Hitler: Evolutionary Ethics, Eugenics, and Racism in Germany*. New York: Palgrave Macmillan, 2004.

Weisbord, Robert G. *Genocide: Birth Control and the Black American*. Westport, Conn., and New York: Greenwood Press and Two Continents Publishing Group, 1975.

Wells, T. Spencer. "Castration in Mental and Nervous Diseases: A Symposium." *American Journal of Medical Sciences* 92 (1886): 455–471.

"Whipping and Castration as Punishments for Crime." *Yale Law Journal* 8 (1899): 371–386.

White, William A. "Sterilization of Criminals." *Journal of the American Institute of Criminal Law and Criminology* 8 (1917): 499–501.

Whitney, Leon. *The Case for Sterilization*. New York: Frederick A. Stokes, 1934.

Willrich, Michael. "The Two Percent Solution: Eugenic Jurisprudence and the Socialization of American Law." *Law and History Review* 16 (1998): 63–111.

"The Winfield Case." *Kansas Medical Journal* (September 8, 1894).

Wong, Caroline. "Chemical Castration: Oregon's Innovative Approach to Sex Offender Rehabilitation, or Unconstitutional Punishment?" *Oregon Law Review* 80 (2001): 1–27.

Woodruff, Charles E. "Climate and Eugenics." *American Breeders Magazine* 1 (1910): 183–185.

——. "Prevention of Degeneration the Only Practical Eugenics." *American Breeders Association Proceedings* 3 (1907): 247–252.

Woods, Frederick Adams. "Some Desiderata in the Science of Eugenics." *American Breeders Association Proceedings* 5 (1909): 248.

Woodside, Moya. *Sterilization in North Carolina: A Sociological and Psychological Study*. Chapel Hill: University of North Carolina Press, 1950.

"Would Limit Voting to Selected Class." *New York Times*, June 7, 1931.

"Would Sterilize the 'Socially Inadequate.'" *New York Times*, January 4, 1923.

Yahya, Harun. *The Disasters Darwinism Brought to Humanity*. Scarborough, Ontario: Al-Attique, n.d.

Zenderland, Leila. *Measuring Minds: Henry Herbert Goddard and the Origins of American Intelligence Testing*. Cambridge: Cambridge University Press, 1998.

Index

About the Author

Mark Largent is a historian of biology and an assistant professor of science policy at James Madison College at Michigan State University. He earned his Ph.D. in 2000 from the University of Minnesota's Program in History of Science and Technology and has taught at American history and history of science courses at Oregon State University and the University of Puget Sound. He is editor of the ABC-Clio *Science and Society* series as well as book review editor for the *Journal of the History of Biology*. His research and teaching focus on the role of American biologists in twentieth-century public affairs, and he has published articles on the evolution/creation debates, the history of evolutionary theory, and the American eugenics movement.